Springer Series in
Experimental Entomology

Thomas A. Miller, Editor

Thomas A. Miller

Insect
Neurophysiological
Techniques

Springer-Verlag
New York Heidelberg Berlin

Thomas A. Miller
Department of Entomology
University of California
Riverside, California 92521
USA

With 148 figures.

Library of Congress Cataloging in Publication Data
Miller, Thomas A
  Insect neurophysiological techniques.

  (Springer series in experimental entomology ; 1)
  Bibliography: p.
  Includes index.
  1. Nervous system—Insects. 2. Insects—Physiology.
3. Neurophysiology—Technique. I. Title. II. Series.
QL495.M54   595.7'01'88028   79-11791

ISBN 0-387-90407-7 Springer-Verlag   New York
ISBN 3-540-90407-7 Springer-Verlag   Berlin   Heidelberg

Series Preface

Insects as a group occupy a middle ground in the biosphere between bacteria and viruses at one extreme, amphibians and mammals at the other. The size and general nature of insects present special problems to the student of entomology. For example, many commercially available instruments are geared to measure in grams, while the forces commonly encountered in studying insects are in the milligram range. Therefore, techniques developed in the study of insects or in those fields concerned with the control of insect pests are often unique.

Methods for measuring things are common to all sciences. Advances sometimes depend more on how something was done than on what was measured; indeed a given field often progresses from one technique to another as new methods are discovered, developed, and modified. Just as often, some of these techniques find their way into the classroom when the problems involved have been sufficiently ironed out to permit students to master the manipulations in a few laboratory periods.

Many specialized techniques are confined to one specific research laboratory. Although methods may be considered commonplace where they are used, in another context even the simplest procedures may save considerable time. It is the purpose of this series (1) to report new developments in methodology, (2) to reveal sources of groups who have dealt with and solved particular entomological problems, and (3) to describe experiments which might be applicable for use in biology laboratory courses.

Thomas A. Miller
Series Editor

Call to Authors

Springer Series in Experimental Entomology will be published in future volumes as contributed chapters. Subjects will be gathered in specific areas to keep volumes cohesive.

Correspondence concerning future volumes of *Springer Series in Experimental Entomology* should be communicated to:

Thomas A. Miller, Editor
Springer Series in Experimental Entomology
Department of Entomology
University of California
Riverside, California 92521
USA

Preface

Much of neurophysiological research is an application of electronics to biology. When compiling technique books on neurophysiology there is a tendency to emphasize the electronics. Indeed, many contemporary technique books begin with electrical circuit theory. I have found these six useful: Brown et al. (1973), Young (1973), Dewhurst (1976), Hoenig and Payne (1973), Geddes and Baker (1975), and Blakeslee (1975).

The treatment presented here emphasizes the biological rather than the electronic. Part I describes instruments, tools, and materials that may be useful in neurobiological research; Part II describes methods used to study unrestrained insects (actographs); Part III describes tethered or harnessed insects; and Part IV describes insect organ and tissue preparations.

Many sources of equipment and supplies are described, and attempts have been made to provide several examples of each item. Where possible, both United States and foreign addresses are used; however, the addresses provided can be consulted directly for information on worldwide distributors. The list of manufacturers and their products will change, but the change will be considerably slower than the half-life for electronic circuitry and semiconductor technology.

In some instances, there are vast differences in prices for certain items. Higher prices are justified if a product is particularly reliable. Although neurophysiologists tend to become attached to certain brands, the author and publishers imply no endorsement of any of the commercial instrumentation described in this volume. The devices described are meant to serve as examples. New products are constantly being developed, and established devices are constantly being improved or modified.

As with any other management decision, the researcher weighs personal construction of devices against the time saved by purchasing the same device from a manufacturer or by having someone construct it. The one point that needs to be stressed is that people have a tendency to underestimate the value of their time. In constructing any device, it is important to subtract not only the total time involved, but also the time which could have been spent at some other aspect of the research. In my experience the most productive arrangements arise when a biological researcher joins forces with an electronics engineer or technician. On rare occasions both abilities are combined in one person.

Background information for this book was gathered over the period from 1973 through 1977. The personal communication citations in the text refer to visits to the laboratories or to the scientists mentioned. Two electrophysiology workshops were held before the writing was completed. One was organized by Professor Dan Shankland and was held in New Orleans in December 1975 with the Entomology Society of America. The second was organized by the author for the International Congress of Entomology and was held in Washington, D.C. in August 1976.

Because of the large number of topics included under the title of neurophysiology, many have been deferred to future volumes. This includes all subjects concerning sensory neurophysiology.

Riverside, California   Thomas A. Miller
October, 1979

Contents

Part I

Materials

Electrodes are the neurophysiologist's test tube, and electrodes for recording extracellular nerve or muscle activity from insects are often an achievement in miniaturization. For this reason insect neurobiologists seek sources for the smallest diameter of wire and must become familiar with materials which might be of use in recording from small tissues.

Glass microelectrodes (micropipette electrodes) for intracellular recording are fairly standard to biology. Microelectrode technology has changed slowly over the past several years in terms of fabrication; however, methods of filling microelectrodes have undergone considerable change recently when Tasaki and co-workers (1968) reported a rapid-filling method using internal glass fibers. This was followed by the development of Omega or other similar glass tubing which is manufactured with a fiber or equivalent already in the tube to greatly facilitate filling. The bother and wait associated with filling microelectrodes is rapidly becoming a thing of the past.

Another modern development is beveling microelectrode tips. Abrasion of the electrode is thought to produce a "syringe needle" point on the glass tip which facilitates cell penetration in some cases. It also may improve the ejection of dye from electrodes for filling single cells.

The physical appearance of electrophysiological preparations has changed little since the Faraday cage was adopted as a shielding device, probably because the problems of vibration and electrical interference are still with us today. However, there is wide variation among individuals in the method of dealing with these obstacles. Some workers prefer using a screen, even to the extent of screening entire rooms, and establishing elaborate earth grounds. Others appear to accomplish the same recording without any screening precautions.

The sections below in Part I describe various materials used in dealing with electrical recording and dissection of insect tissues. While references to many materials might not change in the future, relatively speaking, electronic instruments are constantly changing. Thus model numbers used here will be obsolete fairly soon; however, the names and addresses of manufacturers should be valid for a longer time, and one need only request literature to obtain the most recent information.

Regardless of how thoroughly a technique is described, the individual worker usually settles upon a certain personal modification which is perfected through practice. The sections below contain many different techniques and sometimes several versions are included. These are meant as a guide and a resource. Hopefully they will prove a useful starting point.

1. Glass Pipette Microelectrodes

Extensive references to the preparation of intracellular recording electrodes are now available (Lavallec et al., 1969; Ferris, 1975; Feder, 1968; Geddes, 1972; Glanzman and Glanzman,[1] 1973; Kelly et al., 1975).

a. Sources of Glass

There is an enormous difference in the mechanical and electrical properties of the glass used for making pipettes, not only among the various inside and outside diameters, but also in the composition of the glass itself. In addition, successive orders of glass from the same manufacturer, even the same catalog number, often require realignment of pipette pulling procedures. This is not only costly and time consuming to the investigator, but introduces an extraneous variable into the analysis of data from one preparation to the next. One solution is to restrict stock to thin-walled, small glass tubing, and to order large quantities of glass at each order. The disadvantage of using small tubing is that it lacks the strength of larger stock—a problem which can often be overcome by employing procedures during preparation that minimize tissue penetration by the pipette, for example, desheathing the central nervous system of insects. Microelectrode glass tubing is available as catalog number CSC-25 233001 7740 01 redrawn to 0.032″ O.D. $\pm 0.003″ \times 0.006″$ wall $\pm 0.0015″$ cut 36″ long Specification M-247 from Corning Glass Works. Kimax glass tubing #46485 0.7 to 1.0-mm O.D. and Pyrex glass

[1]Permission to quote freely from Glanzman and Glanzman (1973) was provided by the authors and by the publisher of *Carrier*, David Kopf Instruments.

tubing 0.86 mm O.D., and 0.51 mm I.D. may also be used for microelectrodes. Drummond Scientific (Pearson and Fourtner, 1975) provides glass tubing, as does Garner Glass Co. which specializes in K-G-33 glass, 1.0 ± 0.03 mm I.D., 1.4 ± 0.05 mm O.D. Glass Company of America provides special Microdot® glass as described below and W. Dehn has "theta" glass. Multi-barrel capillary glassware is available from Friedrich & Dimmock and W. Dehn in various configurations. These are suitable for making multiple barrel microelectrodes used in drug studies. W-P.I. has theta and Kwik-fill fiberfilled capillaries.

Friedrich & Dimmock also fill special orders for redrawn glass to suit any specifications. For example, R. K. Josephson (Department of Cellular and Developmental Biology, University of California, Irvine) ordered 1 mm O.D. and 0.13 mm I.D. thick walled pyrex tubing with an internal fiber in place (Omega Dot®) for use in penetrating the cockroach central nervous sheath. The thick walled glass makes a stronger tip which withstands the initial penetration of the sheath. Special fiber-filled glass stock is also available from Clark Biomedical Instruments in England or from K. Hilgenberg, Germany.

b. Microelectrode Pullers

Devices to heat and pull glass tubing suitable for microelectrodes usually provide two stages of tension for the pull, a weak pull initially followed by a strong pull. The strong pull is most often initiated by a microswitch closure once the glass has been pulled to a preset length.

Either vertical pullers or horizontal pullers are common now. Industrial Sciences provides the popular M-1 pipette puller, and Narishige makes the PN-3 (Labtron) horizontal puller. Vertical pullers are supplied by SRI, Narishige (Labtron), and David Kopf (Model 700 C). David Kopf planned to redesign their puller sometime in 1976.

The Industrial Sciences M-1 is popular and, with sufficiently small tubing, can be used for pulling multiple-barreled microelectrodes. The platinum–rhodium ribbon heating filament should be heated to a uniform orange color. Any imperfections in the ribbon heater appear as irregularities of color in the heated filament. The filament needs replacing when it has aged for the sake of consistent microelectrodes. As with any device employing a heated filament, all of the electrical contacts are important and should be checked if irregularities in pulled electrodes are encountered. This is especially true where heater filaments experience excessive oxidation.

Vertical pullers generally use a coil of resistance wire as heater. An advantage here is that a number of tubes may be accommodated in fabricating multiple-barreled microelectrodes since the coil is generally of larger internal diameter than ribbon heaters. Additionally, the chuck holding the multiple glass pipettes of some pullers may be rotated during the weak pull. This manipulation serves to seal the pipette tubes better prior to the strong pull which yields a more uniform electrode. This is not usually possible with horizontal pullers. Glanzman and Glanzman (1973) recommended that the resistance wire coil used with vertical pullers be replaced when the surface becomes pitted from oxidation.

A description of a pneumatic microelectrode puller has been published (Arnold, 1965). Gravity provides the initial slow weak pull; the second strong pull then is provided by air pressure which is actuated via a microswitch on the chuck shaft. Most commercial pullers use electromagnets as the energy source for pulling microelectrodes, and the pneumatic device remains a novelty. Part of the reasons for this are the greater ability to control forces by electricity compared with air pressure

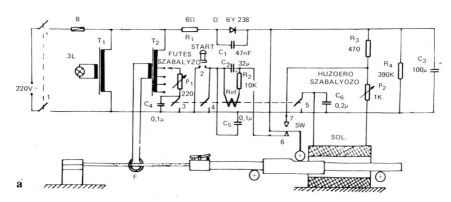

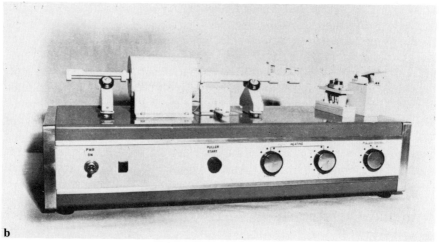

Figure 1-1. Véró: Horizontal microelectrode puller: (a) plans (Fütés szabályzó = heater controls; Húzóerö szabályozó = puller control); (b) photograph.

and the ready availability of standardized electrical power compared to air.

A horizontal puller was manufactured at the Biological Research Institute at Tihany, Hungary. The plans are provided here (Figure 1-1) as an example of the construction involved. A feature not normally provided on microelectrode pullers is some form of cover over the area of the heating element. Air currents in the ordinary laboratory room, even though slight, can have drastic effects on the shape of the electrode tip. For this reason, investigators have found it prudent to develop some simple shield to surround the heater. The advantage is improved consistency in tip shapes produced.

A related procedure is preheating the filament heater before pulling electrodes. This can be done by preventing the chuck from moving during the weak pull period before inserting glass, then pulling electrodes routinely. Alternatively, the filament may be preheated briefly before every electrode is drawn.

Chowdhury (1969) found a blast of air on the microelectrode tip coinciding with the strong pull had a stabilizing effect on the formation of fine tips. Dreyer and Peper (1974a,b) gave construction details for modifying a horizontal puller to include air nozzles mounted near the heater filament and an air valve operated by the shaft of the puller chuck. The cooling air jet allowed more heat to be used, but over a larger range for finer control, and produced tips with a larger cone angle which had improved current passing properties for iontophoresis (Dreyer and Peper, 1974a). Perhaps the greatest advantage was the ability to fabricate uniform high resistance microelectrodes, although no improvement in cell penetration characteristics was noted.

c. Filling Pipette Microelectrodes

(1) Tasaki Method

Tasaki et al. (1968) described a rapid method for filling glass microelectrodes by filling the capillary with three small glass fibers before the pull. Recommended materials are Glass Fibers # 1000, Fiberglass Industries, Inc. This and other developments are described by Glanzman and Glanzman (1973). One of the advantages of fiber filling is strengthened electrode tips for improved penetration of the central nervous sheath for central recording (Pearson and Fourtner, 1975).

Glass capillary tubing may now be purchased with a thin fiber rod already fused inside and suitable for use by the rapid-filling method. This is available from Friedrich & Dimmock or Glass Company of America and saves time required to pack glass tubing with fibers. Frederick Haer & Co. reports that filling prepulled Omega Dot® microelectrodes takes about 1 min when just the end of the shank is placed in the electrolyte solution with the tip up. After the tip fills to the shoulder, electrolyte is then introduced into the shank by a syringe needle to complete back filling.

Upon the introduction of filling solution in the fiber-filled microelectrode by a suitably drawn glass pipette or syringe needle, the microelectrode usually fills to the tip inside of 15 sec due to capillary pressures from the thin fiber strands. Alternatively the shaft of the microelectrode may be held in a drop of filling solution (with the tip up) and the filling solution will be drawn up the microelectrode to fill the tip in a few seconds.

(2) Alcohol

Roger Anwyl (when he was at the Department of Zoology, University of Glasgow, Scotland, 1973) filled by pulling a house vacuum on pipettes with tips down covered with warm undiluted methanol. The methanol was filtered frequently and could be reused if kept clean. Then the electrodes were covered with distilled water for 5 to 10 min, after which the filling solution was injected into the microelectrode in the water. If the shank is filled from a syringe needle, the microelectrodes can be ready in a few hours. If the microelectrodes are immersed in filling solution, they will be ready for use the following day. Electrodes last longer when stored in alcohol than in electrolyte because of corrosion by the concentrated salts.

(3) Boiling

Electrodes have been filled by boiling directly in the filling electrolyte. This is rather drastic since it usually requires a large volume of highly concentrated salt solution.

(4) Vapor Pressure

Filling can be facilitated by the vapor pressure method (Robinson and Scott, 1973). After pulling, the pipettes are mounted tips down in distilled water. A small drop of distilled water is deposited in the neck of the shank near the tip. The microelectrodes may be left covered near a lamp and are filled after a few hours or they may be left overnight for use the following morning, after introduction of electrolyte.

(5) Shell Method

E. J. Ayers (Shell Development Co., Modesto, California 95350) has uncovered a new rapid-filling method which does not require an internal fiber but only involves some pretreatment of the capillary glass. Glass capillaries at least 10 cm long are chosen, cleaned thoroughly in ethanol, rinsed in distilled water, and stored in a jar of distilled water ready for pulling. When electrodes are ready to begin being pulled, a capillary is removed from the jar while a finger is held over the top end to retain about 5 cm of water. The capillary is tipped up, letting the water flow until it becomes positioned in the center. One end of the tube is sealed with a tiny plug of Tackiwax® or other good waxy substance. Next, the capillary is mounted in an electrode puller and pulled. Immediately the micropipette mounted in the moveable chuck (the coolest one) is removed and held at the large end with the tip down at about a 45° angle. The pipette is tapped carefully near the tapered part with a fingernail to make the water droplets flow toward the tip. Once a droplet nears the narrow neck of the electrode, the partial vacuum formed by the rapidly cooling glass will draw the water into the tip. Now, as soon as possible, the held micropipette is removed from the puller and the procedure described above is repeated. It is important to do this while the pipette is still hot, so water will be readily drawn into the fine tip, replacing the air while it does so. Finally the water is replaced by the desired electrolyte.

(6) Removing Air Bubbles

While this technique is rapid, it is not foolproof. That is, some electrodes, after being pulled, retain tiny air bubbles in their tips. Most researchers, it seems, discard such electrodes as unusable, since their tip resistance is usually far above the desired range. This is unfortunate for many, if not most, of these electrodes can be salvaged by using another simple technique. A microscope is needed as well as an electrometer with a three-pole voltage switch (as on those normally used for checking the tip resistance of such electrodes) and a very small electrolyte bath in which to immerse the tip of an electrode. The bath is connected to one end of the circuit and the electrode is connected to the other.

First, a pipette containing bubbles is filled with the desired electrolyte and placed in the electrometer circuit with its tip immersed in the bath. The bath electrolyte should be the same solution as the one in the electrode. The electrode is observed through a microscope erected to provide a clear view of the bubbles in its tip; voltage pulses of about 10 mV then are applied in 1 sec durations and the direction in which the bubbles move is noted.

The polarity necessary to cause an outward flow (toward the tip) of air bubbles from the blocked electrode is determined. Then brief applications of current are applied in this direction. If, instead of being expelled, more or larger bubbles appear, they can often be broken up into smaller bubbles by reversing the flow of current. By again reversing the current, these bubbles can usually be expelled. When this has been accomplished, a previously air-blocked pipette with an excessively high tip resistance often becomes a usable microelectrode.

(7) The Capillary Method

Still another filling method was described by Duling and Berne (1969) and consists of inserting an electrolytically pointed tungsten wire into the barrel of a microelectrode until the sharpened end occupies the area inside the glass tip. This reduces the inner volume and a solution now rapidly fills the tip. Donley (1975) has described some hints on handling and clearing blocked microelectrodes.

(8) Filling Solutions

Filling solutions of glass microelectrodes are standard 3 M KCl, 2 M potassium citrate (Roberts, 1968), or non-chloride electrolytes. One M potassium citrate (Pearson and Fourtner, 1975; Bentley, 1969), 0.6 M (Pearson and Bergman, 1969; Miller, 1971a) or 0.5 M K_2SO_4 (Usherwood, 1973), 1 M potassium propionate (Piek and Mantel, 1970a), 1 M potassium acetate (Kerkut and Walker, 1967; Bentley, 1969; Person and Iles, 1971), and 2 M potassium acetate (Hoyle and Burrows, 1973) have been used for recording from insect nerve or muscle cells. Combinations of K_2SO_4 and KCl used as filling solutions are said to eliminate chloride influx into cells and to stabilize recordings (Lux et al., 1970; Meyer, 1976). Filling solutions can be acidified to improve performance and possibly retard corrosion. Acidifying may also prevent accumulation of microorganisms on the tip. A few drops of citric acid may be added to 2 M potassium citrate or a few drops of HCl may be added to 3 M KCl so as to shift the pH one unit (cf. Wann and Goldsmith, 1972). Bacteriocidal agents have been added to the storing solutions; however, refrigerated microelectrodes filled with electrolyte are not ordinarily kept for more than a week, if that.

Tom Piek (1973, Pharmacology Department, University of Amsterdam, in the Netherlands) filled his microelectrode shanks with 2 M KCl and then covered this with a solid agar/saline mixture with saline over the agar phase. One Ag–AgCl-plated silver wire was inserted into a polyethylene chamber which was fitted over the rear end of the microelectrode and filled with saline solution. The polyethylene prevented damage to the AgCl coating with routine handling, and the saline was not as harsh on the Ag–AgCl electrode compared to 3 M KCl.

Frederick Haer & Co. recommends putting a drop of Copydex or similar latex glue around the top of any microelectrode to prevent evaporation. They further suggest equilibrating microelectrodes by leaving internals in place in the electrolyte for several hours to balance tip potentials and stabilize the surface of the Ag–AgCl internal.

d. Beveling Glass Microelectrodes

Barrett and Graubard (1970) beveled microelectrodes by grinding on a hard circular Arkansas grinding stone with the stone turning at 600 rpm. They were able to attain improved neuron penetration because of the sharper microelectrode tip and were able to infuse Procion dye to stain the same neuron much more efficiently than with nonbeveled electrodes.

Beveling has become widely adapted in a short time because of reported advantages, especially for improved penetration and recording from units in the central nervous system. Several devices were developed for beveling and at least two are being sold. The Model BV-10 was developed from Brown and Flaming (1974) and is being offered by David Kopf. Narishige developed a beveler, the microgrinder diamond wheel or the EG-5 microgrinder, and both are available through Labtron. The Model BV-10, K. T. Brown type micropipette beveler, distributed by David Kopf Instruments measures microelectrode resistance during beveling. W-P.I. announced a Model 1200 beveler under development (summer, 1976).

While the commercial devices provide an immediate guaranteed beveling capability, several laboratories have homemade versions. Most of these use a 60 rpm turntable and various abrasive surfaces (Kripke and Ogden, 1974).

Brown and Flaming (1974) embedded alumina particles on a polyurethane film laid on glass. Holt (Gerry Holt, USDA, Fargo, North Dakota) and Chang used abrasive paper (Thomas Apparatus, No. 6777-J58 0.3 Micron Abrasive film). Still others employ a lapidary diamond compound.

The following description of the Chang beveler (Figure 1-2) was provided by Joseph Chang and Alan Gelperin (Biology Department, Princeton University, Princeton, New Jersey).

The Chang beveler consists of a motorized turntable and a simple lever arm. The lever arm is balanced on the edge of a razor blade and manipulated by a simple rack and pinion control.

First, the lever arm mounting is moved laterally away from the turntable. Then a microelectrode is inserted on the lower end of the lever arm. The lever arm is balanced by moving foil strips on the distal end of the lever arm until the microelectrode tip lies 2 to 3 mm below the plane of the turntable. Then the microelectrode tip is raised above the plane of the turntable by adjusting the opposite end of the lever arm down with the rack and pinion mechanism. With the tip raised, the entire lever arm mounting is moved laterally once again until the microelectrode is posi-

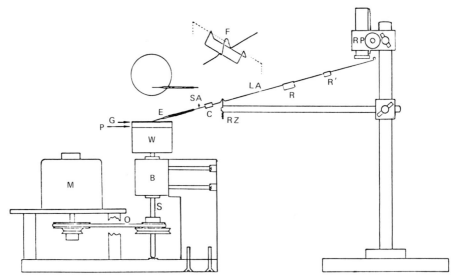

Figure 1-2. Chang microelectrode beveler. B, reamed sleeve bearing (1/4 in.); C, counter weight; E, micropipette, pulled glass capillary; F, detail of the fulcrum; G, grinding surface (0.34 corundacoated microtome grinding sheet from A. H. Thomas Co.); LA, longer arm of balance (whole arm made of 26-gauge wire, stainless steel, or Nichrome); M, motor (any, between 20 and 200 rpm, 50 and 60 rpm best); O, O-ring belt; P, plastic disc, top finished very flat and perpendicular to S; R,R' rider weights, folded aluminum foils; RP, rack and pinion; RZ, razor blade; S, shaft (1/4-in. drill rod) with a steel ball at bottom center, riding on a hardened metal plate; SA, shorter arm of balance, bent very slightly to hold micropipette; W, brass wheel, to give stability by weight. (After Chang, 1975.)

tioned back over the turntable and off center slightly (Figure 1-2, inset). Now with the turntable on (60 rpm) the lever arm is manipulated until the microelectrode tip lies on the abrasive surface of the turntable for 10 to 15 sec.

The microelectrode may be beveled a second time by twisting the microelectrode about 30° on the lever arm shaft and repeating the procedure.

This simple procedure may be modified further by beveling in a Petri dish containing saline solution and thus allowing a constant monitoring of the microelectrode resistance. However, for routine purposes monitoring resistance may be more trouble than is necessary since the entire procedure should not take more than 1 min. Also the force of the microelectrode tip on the abrasive surface will be determined by the original balance point and may be altered accordingly.

A variation on the turntable beveler, the spinning rod, was described by Barrett and Whitlock (1973) (Figure 1-3). In this case the abrasive surface was Schuller 100,000 mesh diamond compound (lapidary industry or metallographic supplies) or Metadi® diamond compound (Scientific Products).

Diamond particles become lodged in the microelectrode tip, but these have no apparent effect on the electrode performance. However, positive air pressure may be applied during grinding to minimize tip clogging. Grinding speeds and times vary and vibration of the spinning rod may be minimized by grinding near a point of support.

An even simpler technique for beveling microelectrodes was recently described by Ogden et al. (1978). The method, termed the jet stream microbeveler, employs a suspension of about two heaping tablespoons of alumina micropolish in a liter of saline solution which is stirred by a magnetic stirrer and forced under pressure through an orifice. The stream is directed at the tip of a microelectrode at about a 45° angle. Beveling is completed in a few minutes by this technique. The authors cautioned that after beveling was completed, exposure to air caused tip clogging.

e. Microelectrode Mounting

Cook and Reinecke (1973) described a floating microelectrode used to record potentials from the hindgut of *Leucophaea maderae* (Figure 1-4). The electrode consisted of a thin silver wire (100–120-μm diameter) chlorided and fed through polyethylene tubing. The thin polyethylene

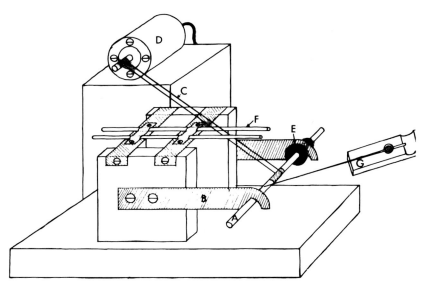

Figure 1-3. Microelectrode beveler (from Barrett and Whitlock, 1973). A, quartz rod; B, stainless-steel supports; C, thread belt; D, motor; E, guide rings; F, thread guides; G, electrode.

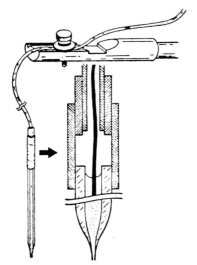

Figure 1-4. Cook hanging microelectrode mount (Cook and Reinecke, 1973).

tubing (Clay Adams P.E. series Intramedic® tubing, or equivalent) was flexible enough to allow the microelectrode to move with the gut.

A drawn plastic tube was described for use as a floating microelectrode mounting by Josephson et al. (1975) (Figure 1-5). Heating the plastic tubing requires practice. Too much heat destroys the plastic and too little heat prevents drawing (see also Gelperin suction electrode, Section 2). Klaus Richter (Animal Physiology Institute, University of Jena, 69 Jena, East Germany) uses a simple glass tube of a few millimeters diameter filled with 3 M KCl. The microelectrode is inserted through a rubber diaphragm in one end and an Ag–AgCl electrode is inserted through

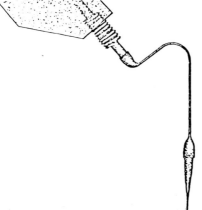

Figure 1-5. Josephson microelectrode mount for a floating microelectrode (Josephson et al., 1975).

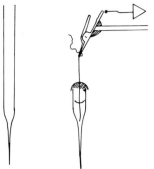

Figure 1-6. Lang microelectrode mount (Lang, 1972).

the opposite end. The glass tube is then held by a manipulator which can be adopted from a simple stereomicroscope rack and pinion mechanism.

Tom Piek (Pharmacology Department, University of Amsterdam, The Netherlands) held his microelectrodes in a tissue clamp connected to a universal ball joint (both are available from medical equipment manufacturers such as Harvard Apparatus or C. F. Palmer). Piek modified his clamp with Teflon pieces bolted to the clamp jaws so that no metal contacted the microelectrode holder.

A hanging microelectrode arrangement was described by Fred Lang (1972) for *Limulus* heart recording and is applicable to insect recording (Figure 1-6). This consists of 1 mil tungsten or silver or platinum wire sealed into the microelectrode tip. At first, scoring and breaking a microelectrode tip might seem drastic and damaging; however, this rarely causes tip breakage as long as the tip is not touched during the break. Some workers wax manipulator probes to the hanging microelectrode until the tip has been guided to the tissue of interest. Once implanted, a hot iron or heated metal probe brought near the wax melts the union freeing the microelectrode tip from the manipulator probe without causing undue disturbance. Another hanging electrode was described by Omura (1970).

The W-P. I. Company sells electrode holders designed for their own amplifier input stages. The holders are simple plastic chambers with Ag–AgCl pellets in one end and rubber diaphragms which hold the microelectrode. The pellet end contains a plug which connects directly to the amplifier probe.

Josephson and Donaldson (R. K. Josephson and P. L. Donaldson, personal communication at the Electrophysiology work shop, XVth International Congress of Entomology, Washington, D.C., August 22, 1976) use a machined plastic block for mounting microelectrodes (Figure 1-7). The silicon rubber gasket was made by allowing silicone rubber compound to

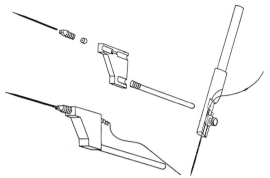

Figure 1-7. Josephson microelectrode mount (Josephson, personal communication, 1976).

cure in a piece of plastic tubing of appropriate internal diameter. The formed gasket is cut to length and perforated with a pin to make the center hole. Alternatively, pieces of appropriate silicone rubber tubing may be sectioned to make gaskets.

The Josephson–Donaldson holder has the advantage of providing a surface for mounting resistors if these need to be close to the microelectrode to avoid interference; or parts may be added to provide pressure to the microelectrode internal for injecting materials from the electrode. In addition, the holder may be mounted at 45 or 90° to the microelectrode which is an advantage for certain manipulators (e.g., for aus Jena manipulators). Finally, it would be possible to remove the Ag–AgCl electrode internal from the Josephson–Donaldson holder with minimal interference to the preparation should a chlorided wire become noisy during an experiment.

2. Suction Electrodes

Suction electrodes were developed independently by several groups. The original idea may have come from sensory hair recording as first perfected by Hodgson and others (Hodgson and Roeder, 1956). The basic idea is straightforward. A nerve bundle is pulled into a tube filled with saline either from the side or from a cut end of an axon. The fit of the nerve must be snug enough to insulate the inside solution of the tube from the surrounding medium bathing the remainder of the nerve bundle. Nervous activity is then recorded from the inside of the tube versus the outside. Ag–AgCl wires usually give records with less noise compared to bare silver wires. The advantage of suction electrode recording or stimulating is that the nerves need not be pulled out of the saline solution and may be left *in situ* submerged in saline solution or hemolymph.

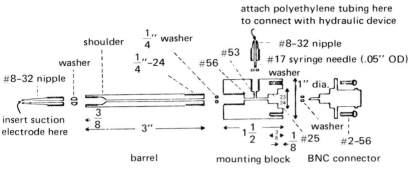

a

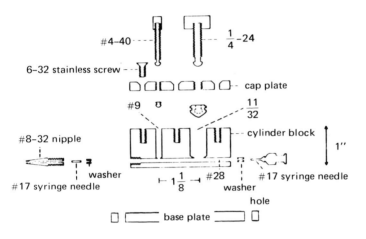

b

Figure 1-8. Florey and Kriebel (1966) suction electrode. (a) View of suction electrode holder, showing the components and their measurements. (b) View of the hydraulic device, showing the components and their dimensions.

One of the early descriptions of suction electrodes was that of Florey and Kriebel (1966) which is reproduced here (Figure 1-8). This version includes a manifold for fingertip control, but is otherwise somewhat awkward if space near the preparation is cramped. A simpler version was described by Delcomyn (1974) employing a disposable syringe and was used for recording from freely walking tethered cockroaches (Delcomyn, 1974) (Figure 1-9). The gasket was stamped from a Neoprene sheet by a #3 cork bore. The internal Ag–AgCl electrode wire was glued into the hole using fast-setting epoxy cement (Epoxi-Patch, Dexter Corp., or "5-minute" Epoxy, Devcon Corp.). This electrode also may be held by the rod shown and requires a holder or manipulator near the preparation.

Robert Josephson and P. L. Donaldson (1976, personal communication, Biology Department, University of California, Irvine,

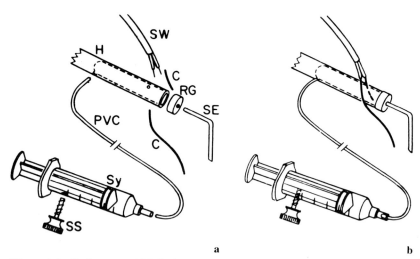

a   b

Figure 1-9. Delcomyn (1974) the suction electrode system: (a) exploded view and (b) assembled view. C, chlorided silver wire (active and indifferent leads); H, suction electrode holder with reservoir (the holder is a 1/4-in.-diameter Plexiglas rod, the reservoir 3/16 in. in diameter, about 1 1/4 in. deep); PVC, PVC 105 size 18 tubing (0.042-in. inner diameter, connecting the syringe to the reservoir, entry via a 5/64-in. hole at rear); RG, rubber gasket (punched from Neoprene with a #3 cork borer, suction electrode hole drilled with #60 bit); SE, suction electrode (0.8–1.0 mm O.D. glass tubing); SS, set screw (10–32, to clamp plunger in position); SW, shielded wire to preamplifiers; SY, 5-ml plastic disposable syringe.

California 92717) recently described a rigid suction electrode made from a disposable 1-ml syringe, but with the suction electrode itself intimately connected with the tip of the syringe needle (Figure 1-10). Such an arrangement could not be used with a delicate preparation where frequent use of the syringe vacuum might be needed. However, this electrode, shown in Figure 1-10, is ideal for simple nerve stimulation and has the advantage of localizing all electrode leads and associated saline to reduce interference or stimulation artifacts.

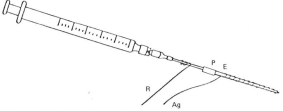

Figure 1-10. Suction electrode construction base on a 1-ml syringe after Josephson and Donaldson (1976, personal communication). R, recording wire soldered to syringe needle; P, plastic tubing; E, glass electrode; Ag, silver reference wire.

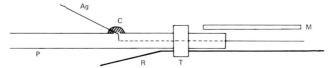

Figure 1-11. Miller suction electrode construction after T. Miller (1974a). The microcap glass tubing is fitted over the protruding 5-mil wire and into the polyethylene tubing. Not to scale. Ag, 5-mil silver recording wire; C, epoxy cement; P, P.E. 60 1-mm-diameter tubing; R, insulated silver reference wire; T, tape collar; M, 10-μl Microcap® tubing.

Another suction electrode (Figure 1-11) was described briefly by T. Miller (1974a). This version consists of a silver wire threaded and epoxy cemented into the end of a P.E. 60 polyethylene tubing (Clay Adams, 0.030″ I.D., 0.048″ O.D.). A 10 ml syringe was used for suction and either a 20-gauge syringe needle fit into the opposite end of the P.E. 60 tubing, or Touhy (Clay Adams) or other syringe tubing adapters may be used. It was found useful to grease the syringe plunger with Vaseline petroleum jelly or high-vacuum silicone grease to prevent air seepage when pulling the nerve into the suction electrode.

The polyethylene tubing was strengthened to receive firepolished 10μl microcaps by holding the end of the tubing close enough to an alcohol flame until the plastic melted and thickened but did not burn. The glass electrodes (10μl size Microcap® capillary tubes, Drummond Scientific Co.) were interchangeable and several sizes could be made for a wide variety of recording conditions, from ventral connectives to lateral cardiac nerve cords.

The Microcap® glass electrodes are made from soft glass and the ends may be collapsed to suitable sizes by manipulating the end near an orange hot nichrome wire (30 gauge nichrome wire, Pelco). The current through the resistance wire may be controlled from a labortory Variac® transformer or similar device (such as solid-state dimmers or heat control rheostats, Ohmitrol PCA 1000, from Ohmite, for example).

It should be noted that heating coils mounted as shown in Figure 1-12 will tend to shift position when heated because of expansion of the metal. Therefore, the heater should be left on to become stabilized for a few seconds before starting. The movement is more pronounced the smaller the number of turns in the resistance wire.

Gelperin (1972) has employed suction electrodes to record nervous activity from the recurrent nerve of the blowfly, *Phormia regina*. These were similar to the Josephson and Donaldson disposable syringe electrode described above, except that a microelectrode was used for the glass portion. A silver wire was introduced into the lumen of a glass micropipette electrode and folded over to run along the outside barrel of the electrode back toward the tip. The electrode was then inserted into a

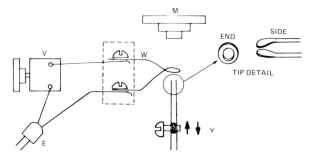

Figure 1-12. Heater for melting Microcap® soft glass suction electrode tips. M, microscope; W, resistance wire; Y, vertical manipulator; V, variac; E, electrical plug. The result of heating the glass tip is shown on the inset in end and side views.

Touhy adapter (#A 1029, 7575-Clay Adams). With a 1 ml syringe fitted, the suction electrode was filled with saline and all air was eliminated. To retain an air-tight hydraulic chamber, a drop of mineral oil was placed on the syringe plunger. Strict elimination of air increased the vacuum transmitted to the mouth of the glass electrode. The electrode tip was cut to the size of the nerve recorded. An indifferent electrode was wrapped around the electrode shaft.

Gelperin also developed plastic suction electrodes. He found that polyvinylchloride (PVC) worked better than polyethylene. He pulled out the PVC tubing by dipping it for a few seconds in hot mineral oil (190 to 200°C) and then pulling rapidly. The appropriate size opening was obtained by trimming the end of the drawn tubing. Again the PVC tubing was connected to the syringe by a Touhy adapter as above. Some prefer to use a chlorided silver wire for suction electrode internals (Delcomyn, 1974); however, this is not entirely necessary since platinum or bare silver would also do, albeit with somewhat increased noise levels.

Some notes on the actual use of suction electrodes may be useful. The lateral cardiac nerve cords of the American cockroach, *Periplaneta americana,* narrow to approximately 25 μm near the rear of the segmental vessels of the heart in the middle abdomen (Miller and Usherwood, 1971). The cardiac cord then widens posteriorly near the juncture with the segmental nerve; therefore, if a suction electrode were applied to the anterior stump, the nerve would be sucked into the electrode until it became stuck at the wider posterior areas. Applying suction on the electrode sometimes filled the inside of the electrode tip with air withdrawn from a tracheole that was present on the lateral cardiac nerve cords. Generally it was most convenient to remove the tracheole before attempting a suction connection; however, in practice this was not always possible. Any air bubbles in the suction electrode tip reduces the efficiency of the vacuum and, if large enough, could interfere with the electrical connection between the nerve and the internal electrode wire thus blocking the recording.

Others have found similar methods for improving the fit of suction electrodes. Don Graham (1973, personal communication), working with *Carausius morosus*, accomplished this in another way by pulling the nerve in to a bifurcation point which sealed and gave a good recording.

3. The Grease Electrode

A more recent innovation for recording and stimulating involves pulling the nerve mechanically into a nonconducting medium (such as mineral oil) for recording or stimulating (Wilkins and Wolfe, 1974). This arrangement has proved highly successful even with very small insect nerves and is referred to here as the grease electrode (Figure 1-13).

The grease electrode can be thought of as a variation on simple extracellular recording electrodes. Recording extracellular activity from nerve bundles by pulling the nerve into a pool of mineral oil (paraffin oil) on metal electrode hooks has been used extensively with mammalian axons. This technique was not used for insects since the mineral oil appeared to shrivel insect axons appreciably (Ken Roeder, 1965, personal communication, Department of Biology, Tufts University, Medford, Massachusets).

More recently Paul Burt (Department of Insecticides and Fungicides, Rothamsted Experimental Station, Harpenden, Herts, England, 1974, personal communication) has been recording from the ventral nerve cord of the American cockroach, *Periplaneta americana*, by using silver wire

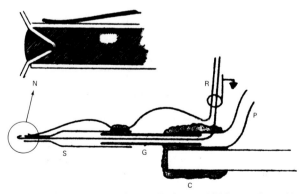

Figure 1-13. Grease electrode from Wilkins and Wolfe (1974). R, wire recording electrodes; P, polyethylene tubing; S, sliding polyethylene tubing; G, glass tube; C, cement and holding rod; The inset shows a nerve, N, pulled into the electrode tip.

hook electrodes with a mixture of Vaseline petroleum jelly and mineral oil spread around the hooked nerve cord to insulate the wires. Burt cautioned that only fresh Vaseline be used, since he noted older discolored Vaseline increased the spontaneous activity of the ventral nerve cord preparation.

We have used the 50:50 vaseline–mineral oil formulation in grease electrodes for stimulating in the hyperneural nerve muscle preparation (Miller and James, 1976). Once in place, the electrode appears to have no detrimental effect on the nerve. Stimulation was carried on for periods of several hours without any appreciable change in the threshold voltage for stimulation.

Pearson (Pearson and Fourtner, 1975; Pearson et al., 1970) has been using Vaseline alone as a basis for extracellular recording from thoracic nerve bundles in the cockroach, *P. americana*. The technique involves pulling the nerve free from saline into the air and then coating the air-exposed nerve bundle with Vaseline (Pearson et al., 1970).

If one electrode hook was used referenced to another electrode in the bath, a monophasic potential was recorded for each nervous impulse (Pearson et al., 1970). If two electrodes were used on the nerve bundle, diphasic patterns are obtained. Kier Pearson cautioned that pulling the nerves firmly could lead to damage and a lack of success in external recording.

4. Silver–Silver Chloride

A further refinement in making electrodes has been the silver–silver chloride coating or pressed pellet. According to Richard Evans (Department of Pharmacology, University of Bristol, England), short lengths of silver wire are cleaned. Silver chloride powder [Mallinckrodt Chemical Works, British Drug House (BDH) or equivalent] is melted in a Pyrex test tube and the silver wires are dipped into the molten AgCl to make an Ag–AgCl electrode. The fused silver chloride in the test tube can be stored in the dark and reused.

For a ground electrode the configuration shown in Figure 1-14 may be used. The electrodes are stored in the refrigerator connected to each other and to ground when not in use (Cooper, 1956). A similar ground electrode with slightly more involved construction is available from Sensorex (as sealed or refillable double junction reference electrodes) or from Transidyne (No. 351 Agar bridge reference electrode).

Randall House [Department of Physiology, Royal (Dick) Veterinary School, University of Edinburgh, Scotland] has used the pressed Ag–AgCl pellet electrode for grounding purposes (cf. Part III) in studies on nerve activity associated with the salivary glands of *Nauphoeta*

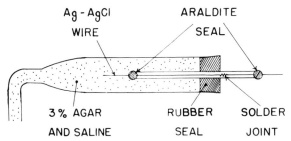

Figure 1-14. Silver–silver chloride reference electrode, suggested construction.

cinerea. The procedure for producing pressed Ag–AgCl electrodes was described originally by Martin et al. (1970) from a U.S. patent (Phipps and Lucchina, 1964). Ag–AgCl pellet electrodes were constructed by Martin et al. (1970) by mixing one part silver powder with two parts silver chloride powder in a piston form. Pressure of 20,000 lb/in.^{2} was exerted on the form to make the pellet. If a silver wire were inserted into the form before pressing, a lead wire would then be left protruding from the Ag–AgCl pellet mass.

House has warned that silver chloride is highly corrosive. Therefore, if the Ag–AgCl pellets are machined, care must be taken to clean the work area or tool surfaces used.

Ag–AgCl pellet electrodes are available commercially from E. W. Wright, Annex Instruments, In Vivo Metrics, and Transidyne General. John Kater (Annex Instruments) has always been willing to construct electrodes to any particular dimension or specification (Figure 1-15). He specializes in nonpolarizable pellet electrodes for electrophysiology. For

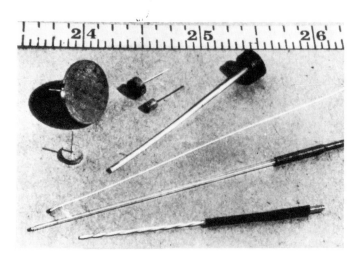

Figure 1-15. Kater silver chloride electrode pellets (Annex Instruments).

best results, it is suggested that a pair of silver chloride pellet electrodes are soaked for 2 to 4 hr with the leads shorted in the saline solution or electrolyte to be used. This will enable the porous matrix of the electrode surface to come into equilibrium with the electrolyte solution. Many workers accomplish this in a laboratory refrigerator which is a handy dark area.

5. Wire Electrodes and Leads

Various forms of wire are used extensively in electrophysiology. Insulated wires are often used for recording extracellular nerve and muscle potentials and shielded cables are used to connect signals from electrodes to amplifiers. As a special case, tungsten is used to make either electrodes or microdissecting probes.

Copper wire is usually the easiest to locate. Silver is also readily available. These can be obtained in large diameters from general laboratory suppliers such as Scientific Products, Fisher Scientific, or Van Waters and Rogers Scientific in the United States.

Wires are also listed as inoculating implements in bacteriology. Silver and other precious metals are available from specialty companies: Martin, Engelhard, Alfa, Sigmund Cohn, Driver-Harris, Kressilk, Reactor, Consolidated Reactive, Johnson & Matthey. Silver, platinum, and tungsten used in electron microscopy as filaments or for evaporating metal coatings are available from electron microscopy suppliers (e.g., Ernst Fullam, Ladd, Pelco, Polysciences). Extremely small tungsten wire or tungsten in a variety of grades of purity and finish are available from the so-called light bulb companies: General Electric, Sylvania, and Firma Osram, for example.

Belden has ultrafine copper wires 47 through 53 gauge insulated with polyurethane, cellulose acetate, or polyestermide films and termed Beldure®, Alenamel®, and Blue Isomid® or Belden ML. Ultrafine wire of precious metal is available from Sigmund Cohn as Teflon-insulated Medwire. Various other companies produce wire in the ultrafine range: Goodfellow, Electrisola, Johnson & Matthey, Engelhard, and The Molecule Wire Corp., for example.

Some electrophysiology suppliers also supply certain types of wires (I.V.M. and Transidyne General, as examples). Silver–silver chloride electrode wire sources were referred to in the preceding section.

Stainless-steel wires are supplied by Goodfellow, Electrisola, and Sigmund Cohn (Medwire). The latter also supplies Teflon insulated stainless-steel wires.

Ordinary single- or multiple-stranded cable or shielded cable are obtainable from standard electronics outlets (Allied or Newark in the United States, Radio Spares in Great Britain, for example), or directly from companies such as Belden or Standard Wire and Cable.

Extremely flexible shielded cables are often useful in electrophysiology, especially where movement is encountered or where stiff cables are a hindrance. Caltron, Cooner, Sigmund Cohn (Medwire), and Popper & Sons (Perflex® cables) are among those companies specializing in this aspect of the biomedical field, and Berk-Tek, Inc. makes microminiature coax cables.

Besides the wires and cables described above, Electrisola provides lac insulating material for coating tungsten wires. The Insl-x Company produces a thinner and insulating lacquer for coating metal wire electrodes, and Diamel® enamel has been used to coat platinum wires (Kay and Coxon, 1956). Medwire provides somewhat expensive Teflon-coated wires including iridium platinum alloys, pure platinum, pure silver, stainless steel, and multistranded stainless steel or silver. Another source for Teflon-coated wires down to 3 mil (76.2 μm) is Omega.

However, since these are thermocouple wires, the types are suited for this application and include iron, constantan, copper, alumel, or chromel. These wires also come as matched pairs of dissimilar metals for thermocouples. Omega also provides uninsulated wires of the same material at diameters of 1 mil (25.4 μm), 2 mil (50.8 μm), and 3 mil (76.2 μm) of 50-ft spools, plus tungsten and tungsten–rhenium alloys starting at diameters of 3 mil (76.2 μm). Omega also has a range of cements and lacquers for insulation.

6. Preparation Area and Accessories

While individual configurations of instruments are included with some of the preparations described in this volume, the elements of a general set-up are detailed here. Intracellular recording procedures for work on vertebrate preparations were described by Dichter (1973).

a. The Room

An ideal situation for research purposes would provide a separate room for each preparation. Since fluorescent lighting commonly interferes with high-impedance electrode recording, it is desirable to switch off the room lights occasionally when it is necessary to trace interference. Even after a preparation is established, it is sometimes necessary to turn off the lighting. For versatility it is better to start with a clear room, with no built-ins with services provided from overhead. This is seldom possible in practice. Access to normal services such as air, gas and vacuum in the standard laboratory are useful.

Rooms which are completely shielded electrically are sometimes designed for electrophysiological laboratories. For the vast majority of preparations, these room are unnecessary. Most shielding, if needed, can be done immediately around the preparation table.

b. Vibration

An advantage of semiisolation of the preparation area concerns the vibration problem. For very delicate work, especially with microelectrodes, slamming doors, cabinets, or constant traffic are a hindrance. Because vibration can be a problem, some workers take extreme measures to stabilize preparations. For example, concrete slabs are occasionally used as the preparation table, or large containers of sand are used as a mounting platform for the preparation platform, or the legs of preparation tables rest in containers of sand.

Parnas (Itzak Parnas, Department of Zoology, Hebrew University, Jerusalem, Israel, 1969, personal communication) filled old sinks with sand and placed marble slabs on the sand and a steel plate on the slab. The sink was isolated in the center of a surrounding desk which held all supporting accessories including the stereoscopic microscope.

In general, problems of vibration are also encountered when positioning a microbalance or an ultramicrotome. Probably the most obvious precaution in choosing a vibration-free site would be to avoid air currents or areas adjacent to hoods, whether in the same room or on the opposite side of a wall in the next laboratory. The same precaution holds for other motorized devices such as centrifuges, stirrers, vacuum pumps, or generators. As a general rule, if an ultramicrotome performs satisfactorily in a certain location, microelectrode work will also be possible there.

c. The Table

For recording with microelectrodes, single isolated tables are generally preferred. Some design and build their own tables; others use common laboratory furniture. Labtron supplies a metal table and Faraday cage. The preparation table shown in Figure 1-16 is presented only as an example. It was constructed of welded steel angle, with wheels for versatility, and construction was begun after a colorlith stone slab top was obtained. Holes were drilled in the frame to mount clamps, accessories, or instruments, and the table was painted to prevent rusting.

Alternatively, tables may be constructed of slotted angle (Dexion®) which has holes suitable for use in mounting. Dexion tables may be constructed to any size, using bolts rather than welding. Dexion is also an excellent source of material to construct tailor-made racks for instruments.

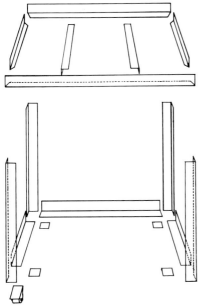

Figure 1-16. Electrophysiology table construction shown in exploded 3/4 frontal view. Dimensions are left out and only one caster wheel is shown. The colorlith top is omitted.

Slotted angle is available from a variety of manufacturers; two United States sources are Lyon Metal Products and Acme Slotted Angle.

d. The Plate

Plates are usually either aluminum or steel. Care should be taken to ensure that the metal is of machineable quality. For example, most aluminum is machineable; however, Type 6061 T6 also heliarcs well and is softer and slightly more amenable to cutting. A list of aluminum alloys is given in Table 1-1 including uses. Aluminum stocks held by distributors usually have alloy, temper, and gauge figures stamped on the material. Types 2000 and 7000 are most often used for plate and 2000 and 7000 are somewhat harder than 6000.

The advantage of aluminum is its resistance to corrosion. A disadvantage is its inability to hold magnetic-based tools and its inability to concentrate magnetic lines of flux to provide freedom from some forms of electrical interference. However, small steel platforms may be used on aluminum plates as bases for magnetic tools. The steel plate corrodes, but this may be retarded by painting and periodic touchup. Plates may be prepared ahead of time by drilling and tapping holes of a standard size for mounting.

Table 1-1. Types of Aluminum and Uses[a]

Series number	Metal alloy	Major use
1000	(99% pure Al)	—
1100	—	Electrical, lighter than copper
2000[b]	Cu	Hard, aircraft construction
3000	Mn	
4000	Si	
5000	Mg	Marine, corrosion resistant
6000[b]	Mg + Si	Soft, shower doors, framing, shelves, easier to extrude and cheaper
7000[b]	Zn	Hard, aircraft construction

[a]Source: Alcoa, Aluminum Company of America, 1145 Wilshire Boulevard, Los Angeles, California.
[b]Most common types used for electrophysiology plates.

The holes may be made in a uniform pattern, in an asymmetric pattern, or not drilled at all. Narishige (Labtron, Appendix) supplies a steel plate, drilled and tapped to hold manipulators.

e. Mechanical Accessories

Some of the standard mechanical accessories which might be useful in the electrophysiology laboratory are shown in Figure 1-17. Figures 1-18 and 1-20 show preparations using typical accessories with aluminum plates. The assortment of pieces in Figure 1-17 are Narishige B-2 ball and socket clamp (a), and rod clamp (b); Harvard Apparatus I-V clamp (e) and No. 208 swivel offset clamp (f); Palmer rod clamp (d); and standard American laboratory rod clamps (h) and thermometer holder (c).

Some of these accessories are included in the general microelectrode setup (Figure 1-18) which shows a ground lead calibrator (a) with ground wire lead (m); a Narishige ball and socket clamp (b) holding the end of a fiber optic (c) light guide; an Olympus zoom stereomicroscope on a universal stand; and the preamplifier probe of a Winston 1090 microelectrode amplifier (e). A microelectrode (f) is held by a Narishige holder (g) and MM-3 micromanipulator (h). The tissue in the perfusion chamber is bathed with saline delivered by a sterile I.V. Set (McGaw Laboratories) including the delivery tube (i) and pinch valve (j) and leading to the drip chamber and saline supply not shown. Saline is continually removed by a vacuum or aspirator line (k). The entire preparation rests on an aluminum plate.

Only a few of the more common mechanical accessories are shown for illustration. The catalogs of Harvard Apparatus, Narishige, Palmer, and

Figure 1-17. Assorted clamps. See text for description.

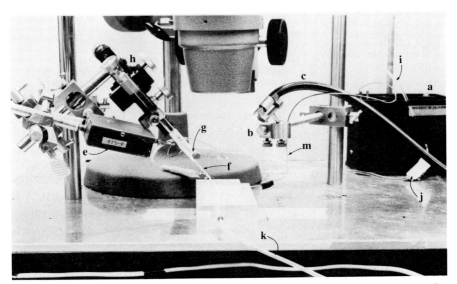

Figure 1-18. Microelectrode instrumentation and recording setup. See text for description.

general laboratory suppliers are recommended to obtain a truer picture of the variety of standard clamps and devices.

Micromanipulators and other rack and pinion devices may also be obtained from aus Jena, Brinkmann, David Kopf, Sobotka, Lafayette, Prior, Leitz, Stoelting, and Burkhard to name a few. Lever-operated positioners which are used in the electronics industry are sometimes adaptable to life sciences research. R. I. Ltd. and The Micromanipulator Company are two examples of the latter. Another source of micropositioners are companies specializing in optical apparatus, and these usually are calibrated in the submicron range (e.g., Ardel Kinametic), and tool companies (Titan).

A host of minor mechanical aids can be found in other places: for example, flexible multi-purpose holders (Flex-set, Wilk Instruments; Spare Parts, Inc.; even Watkins and Doncaster).

f. Perfusion Chambers

Among the most widely used materials for constructing perfusion chambers are the acrylic plastics (e.g., Figure 1-18). There are different types of transparent acrylic sheeting available from which to choose. A recent polycarbonate resin is General Electric Lexan® which has exceptional strength and abrasion-resistant properties, does not readily shatter, and therefore should machine easily. Lexan can be sealed with (1) Silicone Construction Sealants (SCS-1200 Series with S.C.P.-3154 conditioner (General Electric, Silicone Products), (2) Dow Corning® 781 Building Sealant with Surface Conditioner® A (Dow Corning Corp.), (3) Lasto-Meric-Liquid Polymer Sealant (2 part, Polysulfide) (Tremco Manufacturing Co.), and (4) Weatherban Building Sealer (2 part, Polysulfide) (3M Co.).

Plastics may also be sealed with a specific plastics glue made by dissolving shavings from the same material in an appropriate solvent. Ethylacetate or ethylene dichloride has been used successfully in this regard; however, the latter is much more noxious. Chloroform has also been used as plastic solvent for glue but is much less suitable owing to its greater rate of evaporation. Chloroform also tends to attack the acrylic surface somewhat, creating fractures.

Acrylic resins are transparent and are not attacked by nonoxidizing acids, weak alkalies, food oils, or petroleum lubricants (Arnold, 1968). They are attacked by acetone, gasoline, benzene, lower alcohols, phenols, chlorinated hydrocarbons (i.e., chloroform and carbon tetrachloride), lacquer thinner, some esters, and some ketones. Acrylics may be cleaned by water and mild detergent, but not by highly alkaline detergents or window cleaners. Grease and oil may be removed by hexane, isopropyl alcohol, butyl cellosolve, or naphtha. Minor scratches can be removed with a plastic cleaner–polisher or an auto paste wax, or even metal polish (Brasso®, for example).

Antistatic polishers and cleaners are recommended by the manufacturers as listed in the Handbook of Chemistry and Physics (Chemical Rubber Co. Press).

When Plexiglas (Plexiglas, Acrylite, Lucite, and Perspex are trademarks for acrylic plastics) is cut or drilled, localized heating can occur. When the piece of Plexiglas cools, fractures may appear along the cutting surface. This often happens where Plexiglas is drilled and tapped, then pressure exerted by screwing or bolting to the tapped hole. If extensive, such fracturing can lead to leaks around outlet tubes in perfusion chambers or shattered pieces if the walls are thin enough.

One way to reduce heat fracturing caused by drilling is to use an appropriate lubricant to cool the piece being cut. ROCOL® RTD lubricant paste is recommended for tapping and other slow, hard jobs. ROCOL® aerosol is recommended for milling, drilling, and turning. Use of a proper lubricant prolongs turning and extends the life of the cutting tool and the finished product.

Some adhesives are better suited than others for use with plexiglas, and some drug studies are not compatible with plastic. Dan Shankland found, for example, that irrigation of nerve preparations with DDT caused contamination of polyethylene reservoir and tubing (Clay Adams, Intramedic tubing). Trace amounts of DDT ($10^{-10} M$) were enough to produce drastic nervous responses to the presence of otherwise innocuous amounts of cyclodienes (cf. Shankland and Schroeder, 1973; Ryan and Shankland, 1971). Dan Shankland (personal communication) found the trace DDT could be washed away by a slow flush of several hundred milliliters of methanol. After replacing plastic irrigation tubing with glass tubing, no DDT residue accumulated.

Various configurations of perfusion chambers have been developed. The simplest chamber is a wax-filled dish, and this is still a convenient method to work up a new preparation. Sylgard® resin is an alternative material for lining dishes instead of wax. Sylgard® is transparent allowing light to penetrate a tissue from below. It accepts pins readily and is relatively sticky to nervous tissue.

Two problems usually accompany perfusion systems—electrical interference and changing solution levels. The latter can be particularly bothersome. Microelectrode penetrations are often ruined when tissue is moved by changing solution levels; and yet drug studies require replacement of salines. One solution to this problem is described with the sucrose gap preparation (Part IV) and another solution involves special valving (Thomas, 1976).

Plastic chambers may be treated with a silicone coating to reduce adhesion between the bathing medium and the plastic. Siliclad® manu-

Plan view

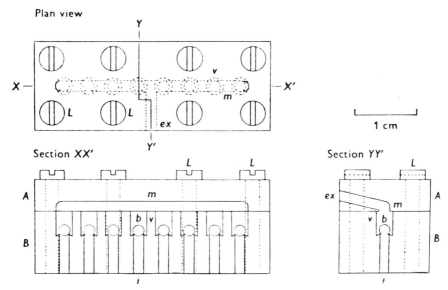

Figure 1-19. A multiway nonreturn valve for physiological experiments (Holden and Sattelle, 1972).

factured by Clay Adams is suitable for this purpose. It renders surfaces hard, water repellent, nonwettable, and acid and organic solvent resistant. Thus, perfusion chambers may be thoroughly cleaned.

The diagram of a valve used for perfusion of drugs in saline was developed by Holden and Sattelle (1972) (Figure 1-19). The valve device is included here since it has general use.

The valve (Figure 1-19) has been built from two machined Perspex blocks (A and B), the apposed surfaces of which have been carefully ground. The upper unit (A) includes a manifold (m) and a single outlet channel (ex); the lower unit (B) contains the valve chambers (v) and their inlets (i). Eight link bolts (L) maintain a firm fit between blocks A and B. Complete isolation of each valve chamber is ensured by a light application of petroleum jelly to the contacting surfaces of the two units prior to tightening the link bolts. Stainless-steel balls are used in the valve chamber and optimal seating is achieved with an inlet tube diameter/ball diameter ratio of approximately 4/5. A modified D-shaped reamer produces a recessed seat for the ball, the sealing action of which is improved by heating the ball before its insertion into the chamber or by gently tapping a center-drilled tool that surmounts the ball in the chamber. Inlet channels (i) and the single outlet (ex) are broached to give tapered holes enabling push-fit connections to stainless-steel hypodermic tubing or nylon tubes.

This valve has been utilized in conjunction with a system of gravity-feed saline reservoirs each linked by tubing to a separate inlet channel of the valve. Opening the tap of one of these reservoirs raises the ball in the corresponding valve chamber allowing fluid to flow into the experimental bath. The valve chambers (v), the manifold (m), and the outlet tube (ex) to the experimental chamber constitute the total dead-space of the system. Rates of flow of 10–20 ml/min have been obtained and tests with colored dyes reveal the apparent absence of leaks. Experiments have, however, been performed to determine the volume of fluid leaking back below the ball in a valve chamber in the nonflow condition when adjacent chambers are in flow. With seven of the inlet tubes connected to the saline reservoirs (35 cm water pressure), a steady rate of leakage of 0–66 μl/min (0–04 ml/hr) is observed in the remaining tube. Under normal operating conditions, the open-ended tube employed to monitor volume changes is replaced by a closed fluid system and it is seldom necessary to have more than one chamber at a time in flow. The device described, which is readily constructed and has a very small dead space, enables a rapid change to any of a number of test solutions without the manipulation of taps close to an experimental chamber. Spheres of various diameters and materials are obtainable from Small Parts, Inc. Methods of maintaining a uniform pressure head in a gravity-feed saline reservoir are described below in Part IV (cf. Figures 4-23 and 4-26).

7. Instrumentation

a. Microscopes

A microscope occupies a central place in the neurophysiological preparation. It is both vital and inconvenient since it is helpful to leave the working area as free from obstruction as possible. These requirements usually call for a microscope to be on a universal stand (also called a swing arm or pillar stand) which puts the microscope stand out of the way. Most dissecting microscopes are available with this feature (American Optical, Bausch and Lomb, Wild, Nikon, Zeiss, Leitz, Olympus).

Dissecting microscopes are useful in magnification ranges from below 10X to over 200X, even over 300X; however, the upper ranges of magnification are limited by shorter working distances. Screw-on lenses commonly reduce working distance without adding detail; that is, the image is larger without a change in numerical aperture (or empty magnification).

The most versatile microscope has stereo and zoom provisions where magnification is continuously adjustable. When it is possible to prepare tissues using lower magnifications at a separate station, then fixed stages can be used with some advantages possible in the type of optical engineering used.

McBain Instruments has recently begun manufacture of a special compound microscope designed expressly for electrophysiology. The unit is

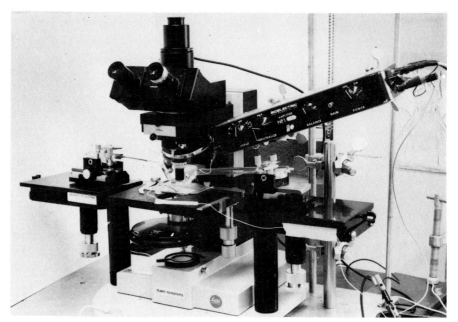

Figure 1-20. Bioelectric NF-1, Aus Jena manipulators and McBain microscope setup for synaptic recording.

fashioned from Zeiss, Leitz, and Reichardt parts to provide Nomarski optics with a rather long working distance (Figure 1-20). Also quite recently, Hoffmann modulation contrast has been introduced by Olympus.

The Nomarski interference contrast system uses a polarizer and modified Wollaston prism above the objective and a second modified Wollaston prism below the condensor with an analyzer. The prisms are correlated so that a source of white light transmitted through a specimen illuminates the field of view in colors of Newton's solar scale. Adjustment of the main prism changes the colors of the object field and brightness changes occur wherever the refractive index or the thickness of the specimen cause a phase displacement of the illuminating beam.

The Hoffmann modulation contrast system does not use a modified Wollaston prism. Instead, condensors and objectives of compound microscopes are modified with a polaroid film to produce interference effects. The result is similar to Nomarski optics but colors are not produced. Only shades of gray show areas where transmitted light has been shifted by a specimen. According to the local Olympus salesman (Los Angeles, September 1976), modified condensors and objectives can be purchased in the range of $800–$1000 U.S.

The descriptions above are of compound microscopes which use *transmitted* light and therefore require specimens which are transparent, or nearly so. Nomarski has been adapted to compound microscopes which use *reflected* illumination. These find use in the microelectronics industry for quality control. The difference between transmitted and reflected optics is significant and units designed for one are not interchangeable for the other.

b. Light Sources

Fiber optics light sources are in extensive use at present for electrophysiology. These offer two clear advantages immediately: intense cold light without heating problems, and lack of electrical interference because the power source and lamp may be housed away from the preparation in a well-shielded cabinet.

There are a number of optics available; some have been fashioned from plastics or glass for particular applications. Many use the quartz halide lamp for intense light. Some require fans to conduct heat from the lamp housing (with possible vibration problems); others radiate heat away. Still other possibilities are DC light sources which present no electrical interference problems. These have been fashioned with automobile headlights and storage batteries.

Some examples of fiber optics are listed below:

1. Fiber-lite high-intensity illuminators: Dolan-Jenner Industries, Inc. Models 140, fixed intensity, and 150, variable intensity tungsten halogen illuminators, and Model 170-D, a fan-cooled quartz iodine lamp.
2. American Optical Corp. Model II-80, a fan-cooled quartz iodine light source; Bausch and Lomb has a similar source.
3. PBL Electro-Optics has Labsource QH 150, QH 100, QH 100-DC, and HG-125; QH 150 is a fan-cooled 150 W quartz halogen lamp illuminator; Labsource QH 100 is a 150 W quartz halogen lamp of fixed moderate intensity without fan cooling. Heat is dissipated by fins on the lamp housing. QH 100 DC provides a DC power source for QH 100. HG 125 is an ultraviolet fiber optic illuminator for use in the far ultraviolet.
4. Zeiss fiber optic illumination system for Zeiss operation microscopes, 150 W lamp.
5. Applied Fiberoptics has a cold light supply, 150 W tungsten halogen lamp.
6. Flexi-Optics Laboratories, Inc. has 106A light source.
7. Narishige LG 96, a microelectrode illumination system (Labtron, Appendix).
8. Rank-Kershaw Fibrox light source (Rank Optics).
9. Circon has an MS 5800 microfiber optic-illuminator with variable intensity. Up to 8000 foot candles of cold, white light are provided.
10. Hacker Instruments 150 W halogen fiber optic illuminator.

8. Electrophysiological Equipment

All devices interposed between recording electrodes and oscilloscopes, cameras, and tape or pen recorders used for visualizing the resultant signals are loosely categorized as electrophysiological instrumentation. This usually includes an amplifier to increase the gain of low-voltage signals or to match high-impedance outputs. Other equipment in addition to the amplifier may be used to perform various manipulations with the signals recorded. This may be counting, signal averaging, differentiating, integrating, or using signals to generate time or amplitude histograms. Contemporary equipment, thanks to microminiaturization electronics, offers a considerable variety of choices at reasonable expense.

Several companies make electrophysiology instruments. References to them are readily available. For example: Federation Proceedings **35**(4), 1090 (1976) lists Dagan, Ealing Palmer, Gilson, Grass, David Kopf, Ortex, Phipps & Bird, Stoelting, and W-P.I. under electrophysiological instruments as having advertised at the FASEB meeting in Anaheim, California, during the April 1976 annual meeting.

In addition to these there are several other companies specializing in electrophysiological instruments such as Frederick Haer, Devices, Digitimer, and ELS.

a. Low-Gain Microelectrode Amplifiers

Since the development of the bridge technique for passing current and recording from a single microelectrode (Fein, 1966), several commercial amplifiers have been developed with this provision. W-P.I. has the M-1 and now M-100 series in various configurations. Bioelectric originally produced a simple microelectrode amplifier, the second version of which, the NF-1, was widely used. Then Bioelectric redesigned their amplifier to the P-1 series which included current-passing circuitry in a modular arrangement. The P-1 and the W-P.I. both have specialized input amplifier stages which mount near the preparation and require special microelectrode mounting. The Bioelectric P-1 series is now offered by Stoelting. The Winston 1090 amplifier and bridge circuit BR-1 also has an input stage which mounts near the preparation and a cable arrangement connecting to the main amplifier; however, the input connector is not specialized. Similar are the ELS system composed of CS-3 and A-4 and the Transidyne neuroprobe (Model 1600).

One limitation of the bridge circuit is the maximum amount of current that can be injected. This is determined by the microelectrode impedance in the first place. Another problem with bridge circuits is an inability to maintain balance conditions after a cell has been penetrated. To be certain of current and voltage values, some investigators still prefer using two electrodes, one to record and the other to pass current.

Other amplifiers are available for microelectrode recording. W-P.I. has the VF-1, essentially a simple solid-state voltage follower with unity gain, negative capacity feedback, and balance adjustment. We have found the VF-1 ideal for extracellularly recorded synaptic currents (Rees, 1974). The Ortec 4661 is a version of the same thing.

Most of the companies supplying electrophysiological equipment provide microelectrode amplifiers. These include Grass, Transidyne, Palmer, Digitimer, Bioelectric (Stoelting), Frederick Haer, ELS, W-P.I., Mentor, and Devices. Some circuit descriptions are published (Cook et al., 1971; Chamberlain et al., 1966; Véró, 1971). An example of a circuit developed by Mr. Charles Roemmele of St. Andrews, Scotland (personal communication, 1976) is shown in Figure 1-21.

b. High-Gain Amplifiers

The requirements for external recording amplifiers are somewhat less stringent than for microelectrode recording. These can be fabricated at considerably less expense. Grass Instruments makes the P-1 series with high-impedance probe. Isleworth manufactures the A-101 which has proven extremely popular. The Biocom 2122 also meets the basic requirements, as does the C. F. Palmer 418/8119 series of amplifiers.

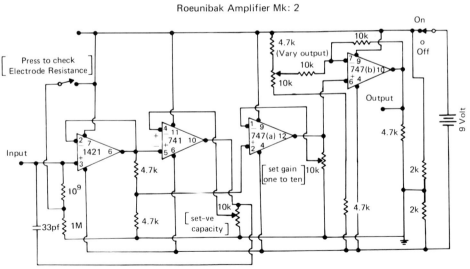

Figure 1-21. Roemmele circuit for microelectrode amplifier (personal communication, 1976).

The Tektronix 122 amplifier has been replaced by the 26A2 differential amplifier which fits into their 2600 modular series. W-P.I. has the DAM-5A differential AC preamplifier with gain up to 1000X.

There are very large numbers of amplifiers available for EEG, ECG, or EMG recordings. In some cases these are also suitable for ordinary recording of nerve–muscle activity. An example is the DATA Model 2124 physiological amplifier, Argonaut LRA 042, W-P.I. DAM-6A AC preamplifier, Phipps and Bird 631 Bioamplifier, Terrasyn Model N-III B, Med Associates MED-110 Bioamplifier, and the Bailey Model LA-1 DC/AC stable low-frequency amplifiers.

The most versatile devices provide high-impedance differential input, switch selectable high gain, low noise, low drift, and a frequency filter for low- and high-frequency ranges and high-input impedance (Mitchell, 1976).

c. Stimulators and Pulse Generators

Simple stimulators for shocking nerves or providing current pulses are now common. Palmer Type 8047 and SRI Type 6020 are listed by Andrew (1972). Harvard, Edco, and Grass have student and research stimulators. Heath has function generators as does Hewlett–Packard, Wavetek, Exact, and Phipps and Bird.

If stimulators have no isolated output, a stimulus isolation unit is necessary to reduce artifact when simultaneously recording from the same preparation. These units are provided separately by Grass, Bioelectric (Stoelting), Tektronix, Digitimer, Ortec, Devices, Palmer, Harvard, Frederic Haer, and Farnell, for example. The Grass SD9 and Devices 2533 stimulators come with an isolation unit as part of the circuitry and are well suited for student and research use. The ELS CS-1 constant current source also does not require a separate isolator.

Optical isolation units have been described (Merrill, 1972a) and an example developed by Peter Buchan is shown in Figure 1-22. There are techniques available to deal with stimulation artifacts (Roby and Lettich, 1975). However, it is important to leave artifacts as part of recorded signals for marking purposes provided these do not hide response signals.

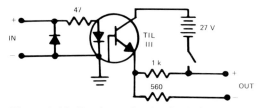

Figure 1-22. Buchan stimulus isolation unit using a Texas Instruments light emitting diode and 27-V DC battery.

Other stimulators including more elaborate programmable devices are available from Grass, Bioelectric, Ortec, BRS/LVE, Frederick Haer, Mentor, W-P.I., and Digitimer.

The Tektronix 160 series pulse generator has been replaced by a 500 series and a 2600 series. The TM 500 series is versatile, but is designed mainly for flexibility in test measurements. The 2600 series includes a nerve chamber, 2601 power supply-main frame, 26G1 rate/ramp generator, 26G3 pulse generator, plus 26A1 operational amplifier board and adapter plug-ins for experimental design. Stimulus isolation is provided by the 2620 stimulus isolator which stands alone. Some pulse circuitry for electrophysiology has been published (Véró, 1972; Lewis, 1967; Atwood and Parnas, 1966).

d. Oscilloscopes, Recorders, and Cameras

A major change in signal handling has been taking place over the past few years in the electrophysiology setup. Pressurized ink recorders have appeared in increasing numbers (cf. Shank and Freeman, 1975, for a recent example). The Brush (Gould) pen system uses thick "ball point pen" ink extruded under pressure from a rectilinear pen drive. With occasional attention to the pen tip, the Brush can operate intermittently for months or years without clogging. The instant-on characteristics of this writing system have given Brush recorders a reputation for reliability.

Both Hewlett–Packard and Société D'instrumentation électronique (Allco) also manufacture pressurized ink recorders similar to the Brush. However, Allco in France built its pressurized ink recorder (Allcoscript®) under a license from Gould and have now been purchased by Gould, so the Brush and the Allcoscript recorders use essentially the same pen system in different packages. The Hewlett–Packard Model 7402A recorder is a separate pressurized-ink design meant to compete with the Brush recorder. The frequency of the Brush recorder is down 3 db at 125 Hz, so that by itself it cannot cover the physiological range beyond muscle potentials. For fast events requiring greater frequency response, FM tape recording has been used more frequently.

FM tape recorders are presently available with a variety of features. The most versatile for physiological measurements are FM tape recorders with a frequency response of DC to a few thousand Hz. Cassette FM tape recorders generally operate at frequencies less than 1 kHz. A few examples of FM tape recorders are listed here with their frequency responses: Data, Inc. 1400, DC to 1250 Hz at 7.5 ips; B & K Instruments 7003, DC to 10 kHz at 15 ips; Philips ANA-LOG 14 or 7 channels, DC to 5 kHz at 15 ips; TEAC/WOLFF R-100, DC to 2500 Hz at 7.5 ips, R-250, DC to 5 kHz at 15 ips, and R-921, DC to 1000 Hz at 3 ips; Dorsch Elektronik 415, 415A, DC to 2 kHz at 7.5 ips; Hewlett–Packard 3960, DC to 5 kHz at 15 ips; Sangamo/Tandberg T1R 100, DC to 2500 Hz at 7.5 ips; and Dallas Instruments FM-4, DC to 2 kHz at 3.75 ips.

The Ampex PR-2200, the Pemtek, Inc. Demco 110, the Sangamo Sabre V, and the Bell and Howell CPR-4010 each have DC to 5 kHz at 15 ips (intermediate band) or DC to 10 kHz (wide band Group I) at 15 ips.

In addition to the above, Data, Inc. describes a Model 1400 B device which may be used with ordinary magnetic tape recorders to cover the frequency range DC to 2–5 kHz at 15/16 ips. It should be obvious from the list shown that FM tape recording at 15 ips has a frequency response of DC to 5000 Hz. If fast events are recorded at 15 ips and then played back into the ordinary pen recorders at slower speeds, e.g., 15/16 ips or even 3.75 ips, the pen records are virtually indistinguishable from those physiological events photographed from the oscilloscope screen. At this reduced speed, any standard pen recorder can be used to record signals. Therefore it would be possible to build an electrophysiology preparation around FM tape recording. The only disadvantage in the long run would be wear and maintenance compared to present methods of data acquisition. However, a good stereo tape recorder capable of two or three speeds is relatively inexpensive compared to most electrophysiology equipment. Only FM tape recorders are very expensive.

The Tektronix dual beam 502A and eight channel 565 oscilloscopes were for several years standard in electrophysiological setups. Grass Instruments used the 502A in their advertisements of the Grass C-4 oscilloscope camera.

While both the 502A and 565 are no longer available from Tektronix, occasionally they are handled by second-hand dealers. Some laboratories are using the Tektronix 5103N as a modern replacement for the 502A. The 5103N should be ordered with MOD 768T modification. This adds three wide-band, direct-coupled amplifiers which provide access to the deflection signals from the cathode ray oscilloscope at rear panel BNC connectors similar to the provision on the 502A. Thus any signal being observed on the oscilloscope is amplified by the appropriate vertical amplifier plug-ins and is available at higher voltages than the original signal at the rear of the main frame. The signals can be used to drive other slave oscilloscopes, or to drive audio amplifiers or recorders, or can be taken to signal processing instruments such as signal averagers, etc. The MOD 768T provision allows signals to be used without disturbing the measuring circuitry at the oscilloscope input.

For neurophysiological applications, the 5A22N vertical amplifier used with the Tektronix 5103N or other 5000 series oscilloscopes is extremely versatile. A low-frequency filter from DC to 10 kHz in six steps and a high-frequency filter from 100 Hz to 1 MHz allow routine signals to be recorded with a selected background. This is especially useful for extracellular recording. Unfortunately the 5A22N is rather expensive and not all of the filtering capability will be used.

Besides the Tektronix oscilloscopes, there are numerous other options available. Philips and Hewlett-Packard have a range of oscilloscopes. Phipps & Bird produce the oscilloscope PB-3 package including bio-amplifier and stimulator for teaching and research. The PB-3 oscilloscope has 50 mV/cm sensitivity. Dumont produces the 5100 oscilloscope with DC to 10 MHz band width and 2 mV/cm vertical sensitivity. Heath/Schlumberger produces the SO-29A, a 50 mV/cm sensitive oscilloscope with a narrow band width of DC to 3.5 kHz which may be limiting in physiological measurements. The Heath/Schlumberger EU-70A oscilloscope is dual trace, 50 mV/cm sensitivity, and DC to 15 MHz band width, and covers the physiological ranges. Digitimer provides a rather complete line of electrophysiology instrumentation including the 3121 oscilloscope. Tektronix has newer low cost T 900 series oscilloscopes with 2 mV/div vertical sensitivity and range DC to 15 MHz.

The following oscilloscope cameras are available: Stoelting's oscilloscope recording camera, Catalog number 12888; Shackman super seven oscilloscope polaroid camera; Tektronix oscilloscope polaroid cameras; Grass Instruments C4 oscilloscope 35 mm continuous recording camera and the Nihon Kohden PC-2A oscilloscope 35 mm continuous recording camera (Lehigh Valley Electronics).

Storage oscilloscopes have been popular since they first appeared. The original versions used specialized CRO tubes which in the Tektronix versions are subject to serious "bleeding." The Hewlett–Packard storage tube is technically superior, but still imperfect. Advance oscilloscopes have just introduced (August 1976) a digital oscilloscope with storage features, but using ordinary CRO tubes. Signals are stored in memory and display by constant renewal. Fast transients can be played back at slow speeds into recorders to accurately reproduce high frequency waveforms. Another new oscilloscope was advertised by Frederick Haer. Its DDO-2 dual-channel oscilloscope has digital display and vertical amplifier outputs. Asher and Merritt (1970) described an extremely useful display technique for use with oscilloscopes having Z axis inputs. This capability should be kept in mind when considering oscilloscope purchases.

e. Signal Averagers and Transient Recorders

Recording and analyzing signals have been made possible by a series of low-cost devices with memory circuits. For single events, the wave form is digitalized and stored until used. This allows very fast events to be drawn out on slow pen recorders by sampling the memory at a slow rate. The Transidyne mnemograph 1500, physical data 512A, Biomation 610 or 8100 Transient waveform recorders function as such signal storage units. The Nicolet Nic-527 signal averager is built around an oscilloscope and appears to be similar to the Advance oscilloscope described above.

Ortec and Digitimer produce a line of instrumentation for signal processing including signal averagers. The Ortec signal averager 4623 uses the memory unit 4620 which can also be used for time and frequency histograms. The Digitimer neurolog NL 750 signal averager can also function as a time histogram instrument.

For a considerably greater investment, the Biomation 102, Nicolet 1070, and Northern Scientific NS 575 represent considerably advanced instruments able to transient store, signal average, perform analyses on the signal such as integration, differentiation, constant addition or subtraction and smoothing plus interface to computers such as the PDP series for additional processing.

For versatile signal handling, several manufacturers offer rather extensive packages. Digitimer has adapted the system developed by Merrill (Merrill, 1972a, b, c) into the neurolog series. A similar versatile arrangement of plug in units is provided by Ortec. These two packages offer reasonably priced signal averagers and histogram capabilities as well as various additional units such as rate meters, counters, pulse generators, and blanks or operational amplifiers for specialized needs, or further adaptation to experimental conditions at the bench. Similarly, Frederick Haer and Transidyne General offer a line of devices, rack-mounted, albeit somewhat less extensive than the Neurolog or Ortec packages. A considerably more involved data processing system is produced by D-MAC which was designed for other applications. The D-MAC system uses Digital Equipment Co. computers of the PDP-8 and newer types. However, these are so specialized that they are of little value unless experiments require complex signal processing of an advanced nature.

9. Dissection Tools

a. Tungsten Microprobe Tools

Tungsten wire may be purchased in 3 in. (76.2 mm) lengths in diameters of 10 or 20 mil (10 mil = 0.010 in. = 254 μm). These wires are cleaned and ready for electrolytic etching and are designed for use as metal electrodes. Tungsten wire may also be obtained as rolls from Ladd Research Industries (20, 25, and 30 mil diameters) and in 10 ft lengths from Ernst Fullam, Inc. (diameters of 5, 20, 25, and 30 mil). Ladd and Fullam are electron microscopy suppliers.

Alfa Inorganic supplies tungsten wire in 0.025 0.127, 0.125, 0.50, and 0.75 mm diameters with a purity of 99.98% based on metallic content. Sylvania and General Electric can supply tungsten wire of varying lengths, diameters, and purity. Merrill and Ainsworth (1972) recommended General Electric type 218 process CLS, 0.005 in. ± 2% diameter, 1.18 in. long.

Tungsten wire may be heated red hot then wiped on a stick of $NaNO_2$ until sharp, or tungsten may be etched by dipping in molten nitrate salt (Garoutte and Lie, 1972). Electrolytic sharpening of tungsten in saturated potassium or sodium nitrite (KNO_2 or $NaNO_2$)(Steel, 1978) is fairly safe for the manufacture of microprobes (although Derek Gammon used NaOH; Cambridge University, ARC Unit, Zoology Department, England, personal communication, 1973). A carbon rod (the carbon from an ordinary lead pencil could serve) is placed in one arm of a glass U-tube and a tungsten wire in the other. The wire and carbon are connected by clips to the secondary of a variable voltage source (Variac® or equivalent) which operates on house voltage (it is advisable to step down 220 V if that is the house voltage). The voltage is adjusted at first to 20 V AC for vigorous etching; then later for finer polishing of the tip, lower voltages are used. To vary the taper obtained, the tungsten metal wire may be dipped into the KNO_2. The rate of dipping will determine the final dimensions. Merrill and Ainsworth (1972) described an etching apparatus plus many tips on ordering, straightening, and use. Narishige (Labtron) offers an Electrolysis grinding machine which is essentially an electrolytic etching apparatus.

The tip shape of etched tungsten wire depends somewhat upon the geometry of the etching chamber. The U-tube described here will direct the etching current symmetrically toward the tip. If a simple beaker were used instead, the tungsten would etch unevenly and produce a flat side nearest the carbon electrode. Naturally, if a flat surface would be useful as a tool in certain applications, then this can be done.

Saturated KNO_2 tends to precipitate and "creep" during drying. This can be prevented at the mouth of the container by coating with grease and closing with cork stoppers between uses.

The finished pointed tungsten wire may now be inserted into a holder (such as wire loop and needle holders, Arnold Horwell; No. 1308-1 loopholder, Pelco; Universal needle holder, Watkins & Doncaster or Narishige W-1 microdissection wooden needle holder, Labtron; Gallenkamp wire holders) and used as a probe for ultrafine dissection. The point may be rounded to a hook by "wiping" the tip against a glass microscope slide or other clean surface. By increasing the angle of incline slightly, a shallow hook is obtained (Figure 1-23) or a longer hook is obtained (Figure 1-23) or even turned back on itself. By applying less pressure, a hook with a correspondingly larger radius may be formed.

Impure tungsten is brittle and tends to shatter or fracture upon cutting cold. The most convenient way to cut tungsten wire into lengths is by

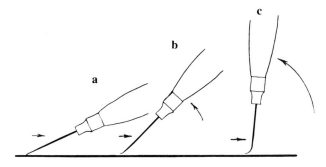

Figure 1-23. Making a hook in a tungsten probe by pressing the probe against a glass side and rotating the holder as shown (arrows).

heating the area to be cut to glowing and then cutting with diagonal cutters immediately upon withdrawal from the flame. Tungsten of high purity is more amenable to cutting.

Merrill and Ainsworth (1972) described annealing and straightening processes, although they recommend the purchase of precut, straightened tungsten wire.

b. Forceps

Either Inox, Dumont, Peer, Royal, Vigor, or Adams brand forceps are suitable for dissection work. These are available from a variety of distributors (Circon, Horwell, Clay Adams, Weiss, Small Parts) and from electron microscopy suppliers (Polysciences, Pelco, Fullam) and are known under the general designation of watchmaker's forceps.

These forceps come in a variety of standardized patterns (Figure 1-24). They may be obtained in carbon steel, stainless steel, or gold-plated carbon (Circon). Fullam offers sharpened Dumont, pattern No. 3. Pelco offers a self-closing, pivot type tweezer, pattern 3, with fine tips which open when pressure is applied and close when pressure is released (Catalog No. 530).

As shown in Figure 1-24, styles 3 and 5 are generally suitable for microdissection work. Some electron microscopy suppliers sell forceps sharpeners; however, fine emory paper may be used to shape old forceps. In fact, tips may be shaped for specific purposes if necessary with a little patience.

c. Scissors

Vannas spring handle microscissors can be obtained from normal outlets (Weiss, Circon, Clay Adams, Roboz). Another type is provided by Down Bros. who also list a 1 mm cutting blade squeeze-action scissors or forceps instrument designed for neurosurgery.

Trident iridectomy scissors are also suitable for larger insect tissues (Model V 38540, Curtin Scientific). A forceps cutter style 15A is provided by Arnold R. Horwell Ltd. Circon makes a variety of

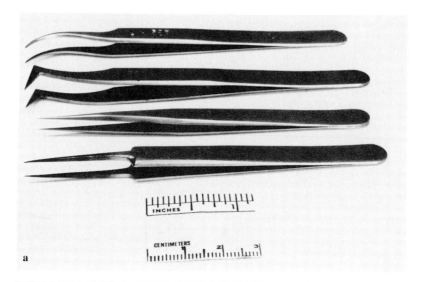

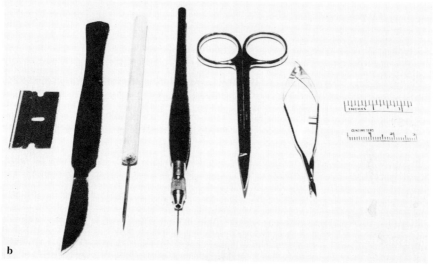

Figure 1-24. (a) An example of styles of forceps available for dissection work. (b) A sample of dissecting tools for work with insect tissues, including iris microscissors and scalpel for working wax.

microsurgical tools such as microknives which may be applicable for insect dissection work.

d. Pins

Insect pins come in a variety of sizes and shapes. Stainless-steel Minuten Nadelns are suitable for microsurgery or pinning tissues (Hamilton Bell Co., tempered block steel pins, sizes "0" to "6", Watkins and Doncaster, stainless-steel headless pins or black steel pins with nylon heads or glass bead heads from 0.010 in. diameter and 10 mm length to size "5"; Clair Armin Karlsbad insect pins, "Elefant" brand, white pins or standard black Imperial Brand, sizes 000 through No. 7 with nylon heads, black or stainless-steel Minutens sizes 0.15 or 0.20 mm only). Other suppliers include BioQuip® and Turtox.

e. Thread and Sutures

Extremely fine thread may be otained by unraveling terylene sewing thread. Virgin silk, stained with methylene blue, is available from Ethicon Ltd. The silk may be unraveled to individual strands which may be used for extremely fine tying. The dye improves contrast immensely. Ethicon also lists 13μm monofilament nylon (Polyamide 6), 11μm nickel chrome wire, and 13μm nickel chrome wire. These materials were designed for suturing in microsurgery. Similar materials may be obtained from hospital supply outlets.

(1) The Veterinary Knot

Tissues can be tied by using the veterinary knot (courtesy of John James, Wellcome Research Laboratories, Berkhamsted, England). The thread is first passed under the tissue, using microforceps to handle the thread (Figure 1-25). The short end of the thread is shown at the bottom of Figure 1-25a. Now wrap the long end of the thread around the right forceps, twice for slippery nylon, once for silk (Figure 1-25, b). Reach over the tissue with the right forcep and pick up the short end of the thread (Figure 1-25c). Holding the short end of the thread, pull the right forcep out through the coils of the long end of the thread (Figure 1-25d). At this point, the thread will usually stay put once released. The two ends of the thread may be pulled tight or left loosely tied to form a loop. The left forcep remains grasping the long end of the thread (Figure 1-25e). The right forcep releases the short end of the thread once the knot is tightened to the desired tension. Now the long end of the thread is again wound around the tip of the right forcep, but *in the opposite direction* compared to the first loop (Figure 1-25e). Now the short end of the thread is again grasped by the right forcep (Figure 1-25f) and the short end of the thread is pulled through the loops in the long end of the thread to make a second knot (Figure 1-25g). This last operation may be repeated again for added strength remembering to again wrap the long end of the thread around the right forcep in the opposite direction compared to the previous tie (Figure 1-25h). Alternating the direction of wrapping as described yields successive square or reef knots in the suture thread (Figure 1-25i); however, the

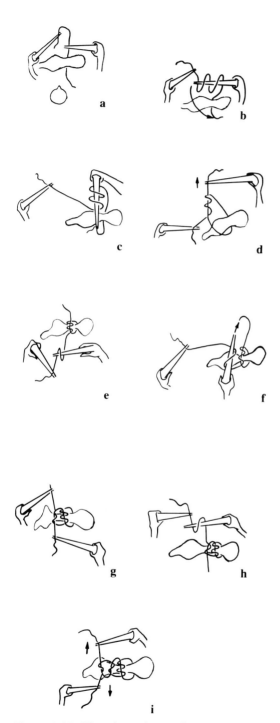

Figure 1-25. Veterinary knot. See text for description.

final tension of the thread around the tissue is determined by the tightness of the first tie (Figure 1-25d). All successive ties serve only to strengthen the knot and prevent slippage. Note that the forcep holding the long end of the thread at the beginning is never released until the entire tying operation is finished.

A convenient way to remember which direction to wrap the long thread around the right forcep is to always lay the right forcep's tip *on* the long thread. Then bring the long thread over the right forcep (Figure 1-25a, e, and h). The successive positions of the long thread alternating away from the operator (Figure 1-25a), then toward the operator (Figure 1-25e), and again away from the operator (Figure 1-25h) after each single tie will automatically provide the correct direction of the wrap.

f. Wax and Plasticine

Materials used as accessories in neurophysiology are sometimes difficult to identify. Therefore, some materials which may be useful are listed and described here.

Plasticine® brand modeling clay has been described in numerous research reports. It holds down appendages or provides a substrate matrix for mounting. Plasticine has also found general use as a convenient holder for tools or electrodes. Plasticine is the registered trademark used for modeling material manufactured by Harbutt's Plasticine, Ltd., Bath, England. Plasticine is distributed in the United States by J. L. Hammett Co.

Tackiwax® manufactured by Cenco has found numerous uses in insect work. Since Tackiwax accepts pins very readily, it is used in dissection dishes, or as the floor material in perfusion chambers. It can also readily stick to housefly cuticle, *Musca domestica,* but not to cockroach cuticle, *Periplaneta americana.*

Dentina ribbon wax (Amalgamated Dental Co.) is similar in consistency to Tackiwax, but Sticky wax (Ash) does not hold together as well as Tackiwax, although it is more rigid.

A 2:1 mixture of beeswax to resin has been used to cement recording electrodes to insect cuticle. The resin is colophonium and is used for violin bows. When insects are under stress, bodily fluids may be exuded over the cuticle from the pore canals. This prevents good wax–cuticle bonding; hence, waxing should be done under anesthesia or other suitable conditions to prevent stress (R. Hustert, personal communication, 1974).

The Dow Corning resin Sylgard® 184 encapsulating resin is a translucent silicone material, "sticky" to nervous tissues. It cures at room temperature, can be used as a floor in preparation chambers, and accepts pins or staples readily. Label instructions suggest adding 10% by weight of Sylgard® 184 curing agent to the resin; the introduction of excess air during mixing should be avoided. Pot life is 2 hr with curing in 24 hr at 25°C; heating will accelerate curing.

g. Lesion Generators

Lesion devices for destroying tissues by heat or localized electrolysis are commercially available (Stoelting, Down Bros., David Kopf) or can be fashioned. The David Kopf RFG-4 lesion generator uses radio frequency for localized heating and was designed for use with small mammals. The Stoelting No. 58040 lesion-producing device provides a current which should be adaptable to work with insects. Also of interest is the Down Bros. bipolar coagulator Mk II for tissue destruction by current in microsurgery. The current is delivered through microforceps.

Vernon (1973) reported the use of such a device for ablating ganglia of Odonata adults. The Vernon lesion generator provides a current to burn the tissue of interest. Insects were attached to a wet copper plate, the cathode, and current was passed to a fine tungsten wire placed in the tissue and used as an anode. The dimensions of the tungsten probe are probably the limiting feature here, although tungsten may be etched to the micrometer range.

Loher (1974) detailed cautery procedures for the pars intercerebralis area of the cricket brain. He used 50μm steel wire electrodes, and pulses of 0.5 to 1-sec duration at 1.5 MHz and 50 mA with Siemens Radiotom 614 coagulator. McCaffery (1976) used a Martin Elektrotom 60 for electrocoagulation.

Grass Instruments makes a current source for generating lesions in small animal preparations. The Grass DCLM 5 is a solid-state device specifically for lesion generation; however, the Grass constant current unit CCU 1 can be used for making direct current lesions. The CCU 1 converts constant voltages to constant currents and may be driven by a suitable pulse generator. A new R.F. lesion generator was described by Unwin (1978) and Steel (1978) employed a homemade device for his microcautery of aphid brain.

10. Microiontophoresis[1]

Microiontophoresis is a method which has been used extensively in studies of drug responses of the vertebrate central nervous system (Curtis, 1964; Kelly et al., 1975). It has been helpful in evaluating neuron responses to putative neurotransmitters. The techniques have also been used with in-

[1]Permission to quote freely from Spencer (1975) was provided by the author and David Kopf Instruments.

sect central nervous studies(Kerkut et al., 1969) and insect neuromuscular studies (Beránek and Miller, 1968; Usherwood, 1969, 1973).

Iontophoresis is the process of ejecting a substance in ionized form, from the tip of a micropipette barrel, by passing a current through an aqueous solution in the electrode containing the substance (i.e., making the anode consist of a solution of acetylcholine, an electrolyte, forces the positively charged acetylcholine out of the electrode tip). For pharmacologically active substances that are not ionized appreciably (bicuculline) a "carrier" ion such as as Cl^- or Na^+ which has a large hydration shell, is iontophoretically ejected and "carries" the desired substance with it. An adjacent recording electrode is used to observe the effect on the cell of the substance being ejected. Larger molecules such as polypeptides are much more difficult to eject from ordinary microelectrodes and only a cumulative effect due to slow diffusion can be measured (H. V. Wheal, 1976, personal communication). However, Brian Brown reports success in readily ejecting proctolin, a pentapeptide, from microelectrodes in iontophoretic studies in locust neuromuscular junction (C. W. Kearns, 1976, personal communication).

Generally speaking, the amount of drug ejected is a linear function of the magnitude of the ejection current used (Curtis, 1964). The amount injected depends on the nature of the substance used, the type of glass, and the nature and size of the electrode tip and is usually consistent if one standardizes these parameters.

For most of the duration of an experiment, it is not desirable to eject the drug except when needed. The application of a retaining current (holding or backing current), opposite in polarity and smaller in magnitude than the ejection current, serves to counter the effect of the osmotic gradient tending to favor diffusion of the substance out of the electrode tip. This is of particular significance where a drug may "desensitize" receptors appreciably.

The magnitude of the backing current is usually of the order of 10 to 20 nA. Unfortunately, prolonged application of the backing current causes diffusion of extracellular ions into the electrode, thereby diluting the drug concentration at the tip. Thus, selection of retaining current magnitude is rather a compromise between unwanted release of the drug and excessive tip dilution as discussed later on. Backing currents can be provided by extremely simple circuitry (Figure 1-26 a).

Ejecting currents are usually as large as are needed to demonstrate an unequivocal effect on the cell under study. Some substances such as the amino acids (GABA, Glutamate) require application currents of less than 20 nA to produce powerful effects; catecholamines, noradrenalin and dopamine, often appear to require currents of 30 to 50 nA, and some of the compounds which antagonize the effects of transmitter substances (bicuculline) often require currents in excess of 100 nA.

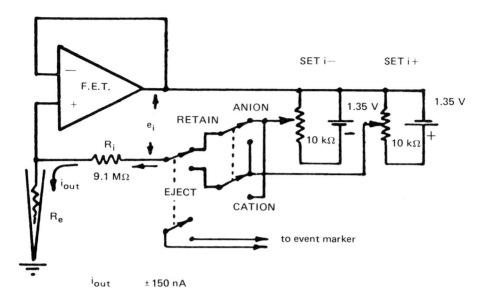

a

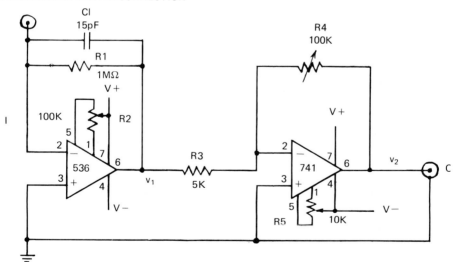

b

Figure 1-26. (a) Constant-current source (Hugh Spencer, personal communication, 1975). (b) Current monitor. The values shown were used for a 10 mV/nA transfer function. (Vincent Salgado, personal communication, 1977).

The amount of current injected through a microelectrode may be monitored by connecting a 100 kΩ resistor in series between the electrode and current source and recording the voltage drop across the resistor with the standard oscilloscope (Beránek and Miller, 1968). Another slightly more accurate method of monitoring the current injected is to construct a current monitor which may be inserted in the ground circuit through which injected current must pass to complete its path (Figure 1-26b).

The current monitor input is most conveniently connected to the bath silver–silver chloride pellet electrode; then its ground is attached to the common ground point of the set-up similar to the connection used for a voltage calibrator.

The circuit shown in Figure 1-26b may be used for either monitoring iontophoretic injection current or for passing current into cells for membrane conductance measurements. It can be calibrated to give an output deflection of 10m V for 1 nA of current injected using the values shown. Resistor R1 may be changed for other current ranges according to the formula $V = -IR_1$. The second-stage amplifier, 741, is in the inverting configuration. Its gain, R4/R3, is adjusted at R4. Offset voltages are nulled with R2 and R5.

Because of the proximity of the electrode tip to the cell membrane, the local concentration of current can produce partial depolarization or hyperpolarization of the neuron. Current controls may be performed to correct for the interfering problem. This is conventionally done by passing a current identical to the one used to eject the drug through an adjacent barrel containing 2 M NaCl using the same time period and procedures as used for the drug. (No holding current is considered necessary for the control barrel.) Responses obtained from cells sensitive to current application are generally discarded unless they are opposite in effect and of much greater magnitude.

There is no need for current controls to be made for all the drugs applied. Only one, the magnitude of the largest ejecting current used, should be adequate. If large ejecting currents of both polarities are used, then two (+ and −) current controls should be performed. For very short ejecting currents, interference with membrane potentials is minimized and correction is probably not necessary.

Multiple barreled glass microelectrodes are conventionally constructed from 1-mm nominal thin-walled pyrex glass tubing either by the "twist" method or by using some form of miniature glass lathe (Spencer, 1973). While the former requires the least outlay and apparatus, the latter produces a more consistent product. However, both techniques are now superceded by the very recent introduction of prefabricated multi-barrel glass tubing (Glass Company of America; Friedrich and Dimmock; Clarke Biomedical Instruments). As a precaution, the capacitance and resistive effects between adjacent electrodes cannot be ignored (cf. Josephson et al., 1975).

While at first glance the cost of this tubing may seem high (U.S. $1.00 per foot on an average), an immense saving in time and energy can be realized—especially when the high failure rate of most electrodes is considered. A one foot length should produce about six usable electrodes. Additionally, the electrode blanks supplied today are often constructed with an integral glass fiber and thus the electrodes should not need to be centrifuged to fill them with the desired substances (Curtis, 1964). This is an especially valuable feature when using substances such as aminophylline or apomorphine which oxidize rapidly. Another feature of electrodes containing an integral fiber is that the drug-containing barrels have a considerably lower electrical resistance, for a given tip diameter, than those made by the other methods, thus permitting smaller tip diameters to be used.

For fashioning multibarrel electrodes by the older glue and twist technique, a large vertical puller of the Narishige type is essential. The bottom chuck of this puller must be free to be rotated manually while the glass is soft, thus ensuring a more efficient seal between the barrels. If the multi-barrel blanks are used, then it is necessary to fire polish both ends (of a 6 cm length) and, following washing with distilled water, the ends should be reinforced with a small amount of quick-set epoxy since the tubing blanks are thin walled and fragile and can be crushed by the puller chucks. Conventional pullers generally cannot handle this tubing because of its diameter (approx. 3 mm), but pullers of the vertical Narishige type can. One could epoxy a smaller capillary inside the center barrel, however, and treat it as a conventional microelectrode, using a puller of the David Kopf 700 type.

If twist electrodes must be filled by centrifugation, a collar of wire may be epoxied to the end of the bundle to retain the electrode in the centrifuge rotor (Spencer, 1975). Inserting glass fibers into multi-barrel blanks is not practicable, for they tend to melt back during the twisting or pulling characteristics of the glass bundle in a rather unpredictable manner.

Once the e electrode has been pulled, the tip can be broken back to about 3 to 6 μm diameter for a seven-barrel tip to permit the electrode to fill and to reduce the resistances of the filled barrel to a manageable value (20 to 100 MΩ for the drug-containing electrodes). This can be done by mounting the electrodes in a carrier of a microscope mechanical stage and ramming the electrode tip into a metal block positioned under the objective. The use of a calibrated eyepiece graticule is essential for this operation. This method gives good control of the tip size. A simple circuit for measuring microelectrode resistance was described by Spencer (1971b).

Prior to filling the electrode, each barrel can be color coded to identify the nature of the drug to be used. Small bottles of modeller's paints are ideal for this and the vials containing the drug solutions (plastic scintillation vials are ideal) should be similarly color coded. Provided care is exercised to exclude dust and provided the microelectrode tubing is washed, blockage due to dirt should not be a problem. Drug solutions should be kept frozen at about -30°C preferably under nitrogen and thawed only for as long as is necessary to fill the electrodes. Similarly, the dry drugs themselves should be stored in the same refrigerator in a desiccator. Very unstable substances (apomorphine, isoprotenerol, etc.) should be made up freshly before use in nitrogen-gassed solvent. The top of a glass serum ampule makes an ideal microliter-volume-mixing container.

Most substances used for iontophoresis must have their pH adjusted either to improve their stability in the case of catecholamines or to become sufficiently ionized to pass current. For example, 1 *M* gamma aminobutyric acid is adjusted to pH 4 (Usherwood, 1973) and 2 *M* monosodium glutamate is adjusted to pH 8.0 (Wheal and Kerkut, 1975). Reference to current literature usually gives pH values for various substances that are iontophoresed. Hydrogen ions are, however, potent excitants and the pH of solutions should almost never be below 3.0.

Once all electrode barrels are filled, their resistance must be checked for open circuit or high-resistance electrodes which might require centrifugation or further tip enlargement to correct. An ideal method is to pass a constant current through the electrode and measure the voltage developed across it (Figure 1-27). Conventional volt–ohm meters and

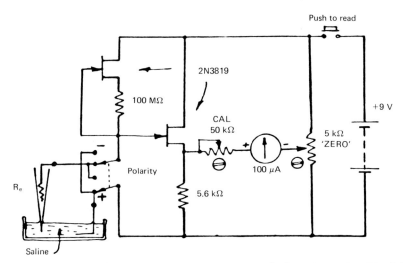

Figure 1-27. Device to measure the impedance of microelectrodes (Hugh Spencer, personal communication, 1975).

VTVMs are not usually adequate, and AC impedance bridges are generally too cumbersome for the rapid type of testing required here. Barrels with resistances in excess of 200 MΩ will probably not function satisfactorily. They tend to become noisy, and, if any drug passes out at all, it will require a high-driving voltage to do so. Such electrodes also tend to increase in resistance after a few minutes in the tissues, possibly due to occlusion by protein on the electrode tip. The current control, if one is used, and extracellular recording electrode barrels containing 2 M NaCl should have a resistance of less than 5 MΩ.

Connecting a multi-barrel electrode to the current source can introduce problems. Provided the electrodes are not filled to less than 3–4 mm from the top, there should be no leakage between the barrels. The lead from the current source to the connector should be a light multi-core-shielded cable, or a bundle of small-diameter coaxial cables.

Electrodes occasionally tend to stop passing current during use (blocking). Sometimes this can be corrected by forcing a high current of the opposite polarity through the electrode, although this is generally only a palliative. Often this blockage is due to a build-up of protein on the electrode tip. Immersion of the electrode tip in a concentrated solution of protease (Sigma pronase) in saline for several minutes can often clean a failing electrode.

In order to reduce the number of variables encountered in iontophoretic drug injection, it is generally accepted that a constant-current source should be used to eject the substances such as a Howland current pump (Dudel, 1975). Constant-current sources, as their name implies, will force a given current through any resistive load, such as a microelectrode, regardless of the changes in the value of that resistance. Real constant-current sources approximate the ideal very well until the voltage developed across the resistance approaches the supply voltage. Since the currents needed for iontophoresis seldom, if ever, need to exceed ±200 nA, the design of a suitable constant-current source is complicated by the leakage characteristics of the semiconductors required to construct the circuits: low leakage field effect transistors (FET) being required to sense current flow. A reliable basic circuit is shown in Figure 1-26a, and is described in greater detail elsewhere (Spencer, 1971a).

Since most available FET operational amplifiers and discrete current source designs have supply voltage limits of about ±20 V, the maximum drive voltage that can be applied across the electrode is of the order of ± 18 V (the compliance voltage) (180 nA into 100 MΩ). For most purposes, this is probably adequate.

Most of the resistance in a microelectrode is concentrated in a very short section immediately before the tip, enclosing a volume of less than 5 μm^3. Thus, even at moderate iontophoretic currents, the current density and power dissipation in this region could conceivably cause accelerated oxidation or hydrolysis of the drugs although at present no such effects have been documented.

The simplest method of controlling drug application is by manually switching between the retain and ejection current sources. Even for six-channel operation, this is often quite adequate; however, when repeated experiments are carried out it can become extremely tedious and a source of error. In testing the efficacy of various substances in blocking or affecting the response to putative transmitter substances such as acetylcholine or glutamate, one accepted technique is to apply the acetylcholine to the target neuron in short (say 10 sec) pulses followed by a 20 sec recovery period between each pulse while simultaneously applying the test substance to the cell for, say, 1 min at a time. This is not particularly convenient to perform manually, and many workers now incorporate some form of "pulser" that will switch the acetylcholine ejection current on and off automatically with relays, while the other channels are manually operated.

Drug release from micropipettes has been shown to be not only a function of the duration intensity of the ejection current, but also of the magnitude and duration of the retaining current (Bradshaw et al., 1973 a, b). If a micropipette remains in the retain mode for any prolonged period, there is a tendency for the drug concentration in the tip of the pipette to be increasingly diluted by ions from the extracellular fluid. On ejection, these ions are the first to be released and the drug ejection rate thereafter increases slowly with time, at a rate dependent on the duration of the previous retain period and the magnitude of the ejecting and retaining currents. Thus, to obtain a consistent response it is advisable to standardize the duration of the ejection and retain periods.

At the present there are two instruments available for iontophoretic injection. One, the Model 160 by W-P.I. Instruments, is a single-channel battery-operated unit with a compliance voltage of ± 100 V and an output current capacity of up to 1000 nA. Several 160s can be operated in parallel for multi-channel operation. It does not have the facility for automatic application of drugs but can be controlled by user-supplied external timing circuits or can be operated manually. It is not suitable for systems involving more than three or four iontophoresis barrels. Because it is battery operated, tying the current return lines together will automatically generate a balance current which can be fed into a balance current electrode. However, as stated before, this is a dubious advantage. Unfortunately, the ejection and retaining currents cannot be preset without the necessity of metering, which can be a disadvantage in some situations.

The other, manufactured by Dagan, Inc., is a six-channel machine. In addition to having presettable current controls which increase the flexibility of operation, the Dagan model 6400 has a built-in pulser and sequential timing circuits to permit serial and/or pulsed application of drugs.

Since computer control of experiments is becoming increasingly common, the 6400 can be interfaced with a standard laboratory computer peripheral system to permit computer control of all parameters of iontophoresis-current intensity, duration, and polarity. Despite the fact that the 6400 is mains operated, the noise injection is equivalent to the battery-operated systems (less than 50 μV per channel).

The output from this circuit can be fed into a suitable strip chart recorder (Heath/Schlumberger) or into a polygraph if resolution is not critical. Counting epoch time is a compromise between sufficient counting resolution to detect depression or an excitation and adequate time resolution to sufficiently display the time course of drug action. Epoch times of 0.5 to 2 sec are usual for fast (greater than 10 spikes per sec) firing rates. For slower firing cells, epoch time of up to 5 sec may have to be used (assuming a full-scale recorder pen deflection for 50 to 100 counts per epoch). An event marker should be present on the chart recorder or one should be fitted to enable the time of drug application to be recorded. The event marker ideally should be driven from the constant-current source eject–retain switch.

If, as is becoming increasingly common, it is desired to examine the effect of iontophoretically applied substances on synaptic transmission, it is necessary to use poststimulus time-interval histogram (PSTH) analysis to examine the before, during, and after effects of the drugs on the cell responses to afferent stimulation. Sometimes interspike-interval analysis can be used to augment PSTH analysis. Both techniques require the use of a special purpose computer (Ortec, Nicolet, CAT 400, etc.), or a general purpose laboratory computer can be programmed to perform these tasks.

The manufacturer provides an optional seventh balance current circuit. One very useful feature of the 6400 is the provision of an electrode-blocking indicator which signals by a light the faulty electrode which is failing to pass current and also signals the event on the event recorder output. When performing repeated automatic runs, this can save many results and much time.

Historically the effects of iontophoretically applied drugs has been evaluated by observing the changes in firing rate of the target neuron, the cell either firing spontaneously or in response to the application of some excitatory substance such as glutamic acid (cf. McLennan and Wheal, 1976). Conventionally this is done by counting the number of spikes which occur during a preset time period (epoch) either digitally or by analog means. This count is then displayed as a voltage on a strip chart recorder. The extracellularly recorded cell firing is amplified and passed into a suitable trigger circuit to differentiate the spikes, and the resultant pulses are fed into the counter. While commercial digital counters and

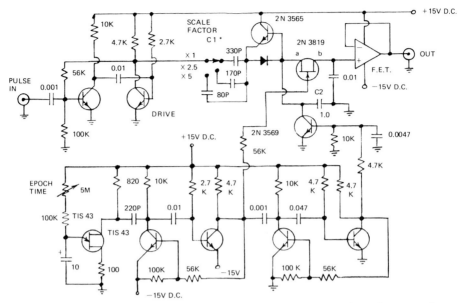

Figure 1-28. Device to convert recorded potentials to pulses for counting frequency of nervous impulses (Hugh Spencer, personal communication, 1975).

digital-to-analog converters can be used for this purpose (e.g., Ortec model 4672 instantaneous frequency/time meter or EKEG model RT 682-L ratemeter), a circuit is given here which will perform the same function (Figure 1-28).

Similarly, the effects of iontophoretically applied drugs on postsynaptic potentials recorded intracellularly can be investigated using averaging techniques. Statistical analysis can be performed on the epochal rate-meter readout, between the firing rate during drug application and during the control periods. For this, a digital record of the number of spikes in each successive epoch is necessary (although with a servo strip chart recorder trace, it is relatively easy to determine the number of counts in each successive epoch). A t test or ψ^2 test can be made between the control and drug responses from this data. However, for most situations this degree of statistical processing is somewhat unnecessary and would be very daunting to perform manually. A response can be considered significantly different if it is $\pm 30\%$ of the amplitude of the preceeding response—as a general rule.

Other iontophoresis circuits and principles were covered recently be Kelly et al. (1975). This includes circuitry, construction suggestions, and helpful references to calibration procedures. Addenda appear periodically (Courtice, 1976).

11. Material Sources

Standard sources of information can be consulted for latest developments
or for addresses and further information. These include: Medical Electronics and Equipment News (MEE News) publishes a dictionary and
buyer's guide, yearly about April. The American Association for Ad-
vancement of Science (AAAS) publishes a yearly guide to scientific instruments as an extra issue of the magazine *Science* on the third Tuesday
in November. The Electronic Industry Telephone Directory (EITD) is a
useful listing of electronics manufacturers. The directory is sold yearly
by the Harris Publishing Co. One March issue of *Federation Proceedings* is devoted to information on scientific products exhibited at the annual Federation of American Societies for Experimental Biology (FASEB)
which is held yearly in April.

Another news sheet is *International Product Digest,* a Chilton international publication whose managing director in Europe is T. Giacomozzi,
6A High Street, Windsor, Berks., SL4 1LE, England, with offices at 201
King of Prussia Road, Radnor, Pennsylvania 19087, USA and C.P.O.
Box 1572, Tokyo, Japan. *Laboratory Equipment* is published monthly by
Gordon Publications, Inc., 20 Community Place, Morristown, New Jersey 07960, USA. Gordon also started *Biomedical Products* (first issue,
April 1976), whose editorial policy includes an invitation to write for further information on companies or products difficult to locate. Their
address is the same as Gordon Publications. Another recent phenomenon is the availability of used equipment. Textronix publishes a list of instruments wanted or for sale in their publication *Tekscope,* Tektronix,
Inc., P.O. Box 500, Beaverton, Oregon, 97005, USA. Lee Lab Supply,
13714 S. Normandie One, Gardena, California, 90247, USA, provides
used amplifiers, recorders, and oscilloscopes, etc. Scientific Resources,
Inc., 3300 Commercial Ave., Northbrook, Illinois, 60062, USA, has used
electron microscopes, spectrophotometers, biomedical instruments, and
general laboratory instrumentation.

One of the most complete listings of companies in the United States is
Thomas Register, an extensive treatment published by the Thomas
Publishing Co., One Penn Plaza, New York, New York, 10001, USA.
The *Register* is in 11 volumes (1974, 64th edition). The first six volumes
list products and the seventh volume contains company credits with
addresses and telephone numbers. Volume 8 contains brand names and
Volumes 9 through 11 are a cross index.

Part II

Actographs: Measurement of Insect Activity

1. Introduction

A device used by Yeager and Swain (1934) to record locomotor activity of cockroaches was described as an "entomotograph," a term suggested by O. E. Tauber. However, this term never caught on, and today devices are usually referred to as one or another type of actograph, a term consisting of a corruption of the words "activity," meaning motion or the state of being active, and the combining form "-graph," meaning something that writes or describes. So actographs are devices for recording activity.

Insect actographs were reviewed by Cloudsley-Thompson (1955, 1961) with references listed up to 1959. Since 1959, several techniques have been developed for use as actographs, and others have been improved. Andrieu (1968) has given a recent valuable review of actographs. He also provides background and results on the application of these methods to the study of grain beetles, termites, etc., *Calandra granaria*, *Reticuliferes flavicolis*, *Zylophoges* and *Rhisophorus* species, plus stridulation and locomotion rhythms of the grasshopper, *Ephippiger ephippiger*. Andrieu (1968) also covered methods for providing signal inputs to drive electronic counter and provided names and addresses for the principal devices used for transducing the primary activity of insects.

The evaluation of biological activity is important for numerous branches of science and critical for agricultural and pharmaceutical fields. Apart from actographic techniques, the evaluation of insect activity has been the subject of reviews that appear periodically (Busvine, 1957; Cloudsley-Thompson, 1955, 1961; Cymborowski, 1969, 1972; Brady,

1969). In addition to these reviews, a few contemporary devices for measuring insect activity are in the series *Experiments in Physiology and Biochemistry*, edited by G. A. Kerkut, and Finger (1972) has recently reviewed psychobiological methods of recording the activity of small mammals primarily, especially rodents.

The selection of an actograph device depends on the size and nature of behavior of the insect, the environmental conditions imposed upon the insect, the physical constraints, and the type of information sought. The type of data accumulated naturally will be determined by the type of actograph selected. Prudent selection of a device minimizes equipment construction and simplifies data handling.

In recent times actographs have been used for the study of circadian rhythms of insects. Normally studies are concerned with overall movement or occurrence of a specific event. Other studies are concerned with the total distance flown or moved by certain insects, or by the time of occurrence of a specific act such as oviposition or feeding, or the circadian rhythm of emergence or courtship.

In designing experiments to measure insect behavior, R. F. Chapman (personal communication, 1973, Centre for Overseas Pest Research, London, England) has emphasized that the insect (locusts in this case) should not see the observer as this introduces complicating parameters into the results. These sentiments were repeated by John Moorhouse (on leave from COPR, London, at Imperial College Field Station, Ascot, England, 1973). Moorhouse et al. (1978) pointed out that for laboratory conditions to be strictly controlled, there must be concern even with the type of lighting used. Since the flicker fusion frequency of locusts is near 100 Hz, presumably fluorescent light discharges would be discernible. Thus, to be overly cautious DC power sources can be used for light.

The extraordinary sensory capabilities of insects cannot be exaggerated. Moorhouse found that his locusts were sensitive to the noise made by movements of some human joints. In fact, Moorhouse's colleague, Ian Forsbrooke, had to stay clear of their behavioral arena during experiments with locusts because of excessively creaky knees.

There are many other examples of elaborate routines used to set the stage for experiments. Preparation of flies for locomotory tests can be quite elaborate, taking up to 4 days (Barton-Browne and Evans, 1960). Dethier (1974) wrote that variability in the proboscis response to chemical stimuli of the blowfly could be explained only by the flies being somehow "different" from one moment to the next. This remained true despite painstaking efforts to rear and handle blowflies under the most identical conditions possible.

There are size and color differences between media-reared and wound-reared screwworm flies, *Cochliomyia hominivoras*, that are not as yet fully understood (Bushland, 1975). Efforts to standardize insect cultures or to compare the behavior of media-reared insects with wild populations are important and continuing (Huettel, 1975; Chambers, 1975).

Although variations in the members of insect populations cannot always be controlled, the conditions of the experiment are much more amenable to control. A good example of oversight was the casual acceptance by many that insects could not "see" red light in the region beyond 650 to 700 nm (covered in more detail in Part II, 3). It is becoming apparent that certain insects can see red light and can establish a circadian rhythm in response to a red light regime.

The design of actographic apparati will ultimately be reduced to the peculiar inclinations of the experimenter (Hammond, 1954). Cloudsley-Thompson (1955) suggested that the simplest type of actograph be used to reduce experimental errors, and that the experimental conditions closely resemble those of natural environments. He further suggested that a particular measurement be reinforced either by direct observation or by measuring activity by an alternative method.

Comparison between the results obtained by different actograph methods has been used to examine seemingly conflicting data arising from studies on locomotory rhythms of cockroaches (Brady, 1969, 1972). The photocell actograph method recorded a decline in the total amount of activity of the American cockroach, *Periplaneta americana,* down to noise level and could not distinguish a periodicity in the remaining activity. As a result, information concerning the periodicity was lost. By way of comparison, the running-wheel actograph (Roberts, 1960) under identical conditions and using the same species of cockroach has recorded very long-term rhythms which did not fade once the experiment was set up (Brady, 1967 a). It has been suggested that the moving substrate of the wheel in the latter method contributed some stress to the cockroaches which reinforced overall activity (Brady, 1967 b, c, 1969). The photocell actograph was not consistent, and it was impossible to compare intensity of activity from one experiment to another. Furthermore, the photocell method did not distinhguish between locomotion, feeding, and preening. If one were interested in cockroach locomotor rhythmicity, then the photocell method could be misleading.

It is clear, then, that different actograph techniques can yield different types of information. Some methods surely contribute something to the conditions perceived by the insect. For example, gross brain damage can make animals hyperactive, masking rhythmicity, and mild stress such as encountered in surgical procedures can also cause enhanced locomotory activity (Brady, 1969).

Certainly any restraints imposed upon the insect have an effect on locomotion performance. This is mentioned in Dutky et al. (1963, p. 281, Discussion) where *Leucophaea maderae* activity when unrestrained was normally very low during the day, while cockroaches with implanted electrodes were far more active during normally quiescent periods such as during photophase.

Our own experiments on restrained house flies show differences in susceptibility to insecticides when compared to unrestrained house flies. In all cases tethered house flies were statistically less susceptible to poisons than those held in cages (Miller, 1976). Similarly, insects treated at different times of the day are more or less susceptible to poisons depending on the time (Batth, 1972), and Bauer (1976) found a decrease in cholinesterase activity in the suboesophageal ganglia of *Schistocerca gregaria* brought about by artificial stimulation of the locusts in a wind tunnel.

Quite apart from biological variability encountered with insects, once an actograph is selected, two methods have been used to record data. Either analog or event signals from actograph transducers have been pen recorded without refinement, or analog activity has been converted to unit events to drive event recorders or counters. The latter perform as relay contact closures or the electronic equivalent.

For completely automatic devices which may be left for hours, days, or weeks, the event recorder (Evans, 1975; Shipton et al., 1959) some form of reset counter-printer are appropriate. Concerning commercial actographs, Stoelting, LKB, or Columbus Instruments provide this capability, and BRS/LVE provides the basic building blocks. Periodic photographs taken of counter dials have been used to record data at intervals from continuing experiments in the absence of a counter-printer (Green and Anderson, 1961). The ideal, of course, would provide an immediate readout of activity at the terminus of the experiment into a publishable form which is also possible (cf. Miller, 1976).

The details of several actographs are given below. Some of these are not available elsewhere and all have peculiar advantages or disadvantages. Some details of the experimental arrangement are usually included, and additional information is given when pertinent.

2. Mechanical Actographs

The original actographs were completely mechanical devices such as the so-called rocking-box or tilting-box actograph or variations of these. An insect container was balanced precisely so that a lever arm attached to the container wrote on a smoked drum as the insect balanced and unbalanced the lever arm. Later, data were collected by making and breaking a contact, often of a metal conductor dipping into a pool of mercury.

The former approach was used by Szymanski (1914) who is credited with being the first to use an actograph to measure insect activity. A rod was balanced on a knife edge with an insect cage and a weight on opposite

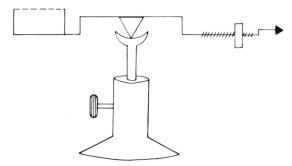

Figure 2-1. Szymanski (1914) box and needle actograph for recording insect activity on smoked paper.

ends of the rod (Figure 2-1). A pointer was attached to the weight and scraped on a revolving smoked drum. Later, Edney (1937) used a similar device to record movements of locusts. This same arrangement, being simple and straightforward, is still being used, especially when an investigator is interested in overnight or preliminary information on behavior (Macauley, 1976, Rothamsted Experimental Station, personal communication). It is ideal, for example, as a preliminary step in setting up more elaborate experiments or apparati and in checking actographic data from other sources or methods.

Hammond (1954) employed a lever and balance actograph principle; however, instead of writing directly on a drum, a vertical slit was attached to the balanced insect container. A light beam was positioned such that the image of the slit was projected onto and recorded by a camera with continuously moving film. The activity of insects weighing 3 mg could be detected, but, not surprisingly, extraneous light and drafts were a problem.

Lutz (1932) was one of the first to develop chart drive and pens to replace smoked-drum kymographs. Lutz evidently made almost everything by hand, including chart drive, pens , and ink supply system. This tradition of the resourceful researcher is evident today (Evans, 1975) as refinements are continually developed.

Versions of the original balance and rock-box devices are still used today even though the kymograph or recorders have changed. Modern event recorders are ideally suited to record the on-off action obtained from rocking-box actographs. Even the more sophisticated modern actographs giving graded responses often have their outputs converted to pulses for counting and this often amounts to another version of the contact closure principle, although modern methods can also record other more sophisticated information.

Brian (1947) pointed out that traditional recording devices provided information about period and intensity of activity without distinguishing the type of activity. For photoperiodic or circadian rhythm studies, this information is sufficient.

a. Running Wheel

The activity wheel, so familiar to the worker studying small animal behavior, has been adopted for studying insect activity. Running wheels constructed by Roberts (1960) for cockroaches consisted of a balsa wood frame and nylon netting. A hexagonal nut mounted on a shaft tripped a microswitch for recording activity. Alternatively, a frictionless switch was constructed in which ferrite rods mounted on the axle of the actograph interrupted an oscillator circuit. By the latter method, the movements of small cockroaches were recorded.

Roberts could leave the actographs and cockroaches for a month by providing a supply of water and a food pellet. The Roberts box was vented by pulling air through a light baffle. This design has proved quite popular for measurements of cockroach activity. The Roberts wheel has been copied and developed by others on numerous occasions.

Lipton and Sutherland (1970a) evaluated the capacitance actograph of Schechter et al. (1963) and decided against using it. They cited the cost involved in duplicating one device for several independent actograph measurements and the sensitivity to electrical interference which evidently was bothersome. The Roberts (1960) running wheel actograph was substituted instead.

Ball (1971) also evaluated several actograph devices for recording cockroach locomotor activity. He too chose the wheel actograph based on Roberts' (1960) design with some modifications. The wheel cage of Ball (1971) was made of a circular plastic container 6 in. in diameter and 1.25 in. deep. The wheel axle was supported by two miniature ball-bearing rings mounted in the wall of a light-tight container (Figure 2-2).

A light baffle mounted outside the wheel box was attached to the end of the wheel axle and this interrupted a light beam. Alternatively the light baffle was replaced by a disc containing small magnets and a reed relay switch in place of the light source. Lipton and Sutherland (1970a) used Hamlin Inc. Model ORS-5 electromagnetic switches and Esterline Angus Model A620X recorders.

Although the magnetic switching arrangement seems to be the simpler of the two detectors on first inspection, Ball found that the photocell system required less maintenance. In addition, if the lamp used for such a detector were operated at less than the rated voltage, its life would be increased.

As a matter of interest, Ball found that locomotor activity of male cockroaches was more consistent than females. He also found some

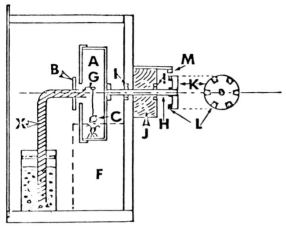

Figure 2-2. Ball (1971) running wheel actograph for use in detecting cockroach activity. A, running wheel; B, baffle; C, food pellet; D, water source; F, shield; G, anchor for wire suspending food pellet; H, axle; I, ball bearing races; J, support block; K, plastic disc holding magnets; L, magnets; M, reed switch; X, plastic tube containing dental wick.

females which were active during the photophase. In general, periods of activity coincided with the interface change in light conditions, at lights on or at the moment of lights off. The least amount of locomotory activity in a 12:12 light–dark regime occurred immediately before light on according to the published records (Ball, 1971).

Sokolove (1975) used running wheels to study cricket, *Teleogryllus commodus* (Walk), locomotory activity. He employed magnetic reed switches (Heathkit, Model N1. GDA-97-1) and one magnet with counterbalances on the wheel shaft.

b. Rocking-box

As mentioned above, contact movement, or rocking-box or tilting-box actographs are among the oldest developed for recording insect movement. A wide variety of chambers based on this principle has been developed. Many experimental arrangements were described by Cloudsley-Thompson (1955, 1961) and Andrieu (1968).

Other examples are given in the review by Cymborowski (1972), and in the works of Green (1964) on the behavior of the blowfly, *Phormia regina,* and Brady (1972) on tsetse fly behavior, *Glossina moristans moristans* West. The boxes of the latter weighed 2 g and could be tripped by a 10-mg weight at either end astride the pivot.

The chambers of Green (1964) are instructive in that they included 0.048 in. (1.22 mm) diameter polyethylene tubes which provided nutrient solutions from microburettes. Individual feeding could be measured to the nearest μl. Contact across a 28 V DC was used to activate simple counters, and accumulated counts were photographed every 30 min. Cymborowski (1973) measured circadian rhythm activity in the house cricket using a tilting-box. The cricket box consisted of a circular tube making an "endless" corridor that was balanced on an axis through one diameter. Mercury switches were used to operate event recorders. Food and water were supplied in the central portion of the toroid-shaped structure out of the runway area.

A variation on the simple rocking-box method was developed by Edwards (1958). Instead of making clean contact with a pool of mercury, Edwards fastened a gelatin capsule to the middle of a length of graphite lead of the type used for replacement fillers in mechanical pencils. The graphite rested on its ends on the sharp edge of razor blades. This resting resistance was measured in a range of a few thousand ohms.

One adult ambrosia beetle, *Trypodendron lineatrum* (Oliv.), in the gelatin capsule caused resistance changes of 10,000 Ω. These changes were amplified by an inexpensive Heathkit phonoamplifier, which in turn drove a relay and counter circuit (Figure 2-3). Interference necessitated careful screening and grounding of the entire apparatus.

Although Hammond (1954) was able to record the activity of insects weighing 3 mg, Powell et al. (1966) did not find the rocking-box method suitable for mosquito adults weighing 1 to 2 mg each. In addition, they considered the testing of large numbers of mosquitoes to be impractical.

c. Flight Mills

Flight mills or roundabouts have been used since at least 1911 (Axenfeld, 1911; Macauley, 1974) and insect flight has been studied by measuring vibrations (Boettiger and Furshpan, 1952) or by stroboscopic methods (Williams and Chadwick, 1943). However, more recent studies on flight actographs based on the flight mill or tethered flight in wind tunnels have been developed in many independent laboratories.

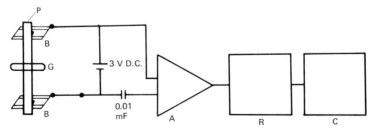

Figure 2-3. Edwards (1958) carbon rod actograph for recording the activity of very small insects. A, amplifier; B, razor blades; C, counter; G, gelatin capsule; P, carbon rod; R, relay.

Flight mills were used in basic physiology studies and originate from early work on locusts by Weis-Fogh (1956) and Krogh and Weis-Fogh (1952). Other studies were concerned with practical information concerning flight range and speed as reported by Hocking (1953). Wind tunnels have been used in conjunction with very elaborate and sensitive transducers, which have been constructed to measure the forces developed by insects during flight (Gewecke, 1975; Camhi, 1970; Reichardt, 1973; Geiger, 1974) in attempts at understanding the neuromuscular basis for flight, optokinetic responses, and turning tendencies.

The use of the flight mill may have generated the broadest amount of practical information concerning insect flight. Data concerned with the duration of flight (Kishaba et al., 1967) and more recently with the propensity to fly (Schroeder et al., 1973) are perhaps the more pertinent parameters concerned.

The subject of flight actographs including details of construction of flight mills is so extensive that it will be described more thoroughly in another volume in this series.

3. Photocell Methods

The photocell or light beam actograph has been extremely popular. A description of an early device by Brown (1959) referred to the Wigglesworth text of 1953 with the statement "insects are generally not sensitive to the red end of the spectrum." This casual acceptance of insects incapable of perceiving red light has carried over, and many authors have assumed that most insects do not perceive red light. The procedures of Cymborowski (1973) are fairly typical. He used a photocell actograph with red light of wave length longer than 700 nm and apparently obtained responses to crickets comparable to results with an actograph of the tilting-box type. Cymborowski, referring to the work of Burkhardt (1964) stated that "the great majority of insects fail to react to wave length within limits from 600 to 650 mμ and over." Dreisig and Nielsen (1971) used a photocell method to measure the activity of the German cockroach, *Blattella germanica*. They simply mentioned a lamp with an infrared filter and obtained data that suggested that the cockroaches were not affected by the presence of the photocell lamp in terms of their resultant locomotory activity.

Godden (1973) found that a red light:dark regime of 12:12 could entrain both egg laying and locomotion activity in the stick insect, *Carausius morosus*. His light was a water-jacketed fluorescent light source (commonly used by Colin Pittendrigh's laboratory, Stanford University, Palo Alto, California 94305) with a Kodak 1A Safelight filter interposed between the fluorescent lights and the insects. Godden (1973) pointed out that the stick insect is unusual among insects in its response to red light (cf. Godden and Goldsmith, 1972).

More recently Sokolove (1975 and personal communication, 1975) has found that crickets *Teleogryllus commodus,* have both locomotory and stridulation rhythms that are entrained by red light. Werner Loher (Entomology Department, University of California, Berkeley, California, 94720) also entrained crickets to an LD cycle using red light. The light source was similar to a Kodak 1A Safelight filter in that it was opaque to wavelengths of light shorter than 600 nm.

Drosophila melanogaster is said to be able to "see" red light, but the circadian rhythm is not affected by a red regime (Frank and Zimmerman, 1969; C. S. Pittendrigh, personal communication, 1975). Similarly the circadian rhythm of the Pink Bollworm, *Pectinophora gossypiella,* cannot be entrained by red light (Pittendrigh et al., 1970). In fact, the total number of insects not responding to red light far outnumbers those for which red light can cause changes in circadian rhythms. Bruce and Minis (1969) found that egg-hatching rhythms could be entrained by blue light but not by red light. Cymborowski (1973) and Brady (1967b) found that while photocell or rocking-box actographs gave different quantitative results for technical reasons, these methods both gave similar results in response to changes in ambient light conditions, again suggesting that the red light used for the photocell detection was not contributing to the behavior of the cricket or cockroach respectively.

Attitudes have changed slowly concerning red light and its perception by insects. A recent communication from T. H. Goldsmith (Biology Department, Yale University, New Haven, Connecticut 06520, October 1975) is instructive.

> "It is quite misleading, indeed incorrect to state that insects cannot see red light. Some, such as cockroaches and honeybees, are probably less sensitive to red, relative to green than are human beings, but the nature of the long wavelength limit of the spectrum is that it has no sharp cutoff. Moreover, there is some indication that at least certain butterflies may be quite sensitive to red light. I would feel that any work using red light to observe locomotion of insects which are presumed to be in "darkness" is highly suspect."

Contemporary reviews give the visible spectrum for insects from the region of 300 nm of the near uv to wavelengths of longer than 600 nm (Goldsmith and Bernard, 1974). Most species of insect are said to be less sensitive to red light than some vertebrates with some important exceptions. Most of the published sensitivity maxima of adult insect compound eyes are in the region of wavelengths shorter than 600 nm with sensitivity falling off rapidly at longer wavelengths.

The device constructed by Brown (1959) using a photocell principle was later improved to use infrared detection (Brown and Unwin, 1961;

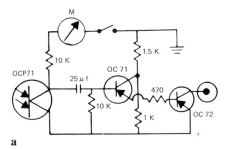

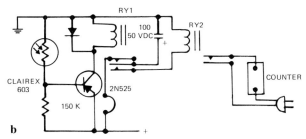

Figure 2-4. Photo actographs: (a) Brown and Unwin (1961); (b) Finger (1972) photo actograph circuits.

Figure 2-4a). The improved version served as the standard reference for photocell circuits for some years. In fact, it is analogous to the work by Roberts (1960) on running-wheels in terms of the number of later studies that used photocell circuits and referred to the Brown and Unwin device. The equivalent circuit using a photoresistance is shown in Figure 2-4b.

Phipps (1963) further modified the Brown and Unwin (1961) detector for use with locusts (Figure 2-5). He covered the lamps with infrared filters (Ilford 207) and designed the red light beams to criss-cross the activity area. Marching activity could be readily discerned from the data, and turning activity, which did not interrupt the light beam, occurred but not often enough to introduce large errors.

Brady (1967b) used the Brown and Unwin (1961) photocell device during studies on the circadian rhythms in cockroaches. An Ilford No. 207 filter was used which provided red light of longer than 720 nm wavelength since *Periplaneta* were considered to be sensitive to red light near 650 nm. As mentioned above, the photocell method of recording the locomotory activity of cockroaches was found to be less useful than the running-wheel method of Roberts (1960). It was difficult to compare peak intensities of activity and, after a week, the locomotory activity was only slightly above the background noise level with the photocell actograph.

The OCP71 photosensitive transistor has also been adapted for use with movement detection on a smaller scale (Pickard and Mill, 1974,

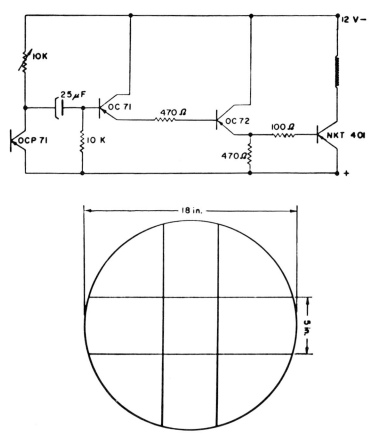

Figure 2-5. Locust photo actograph circuit and chamber (Phipps, 1963).

1975). For this purpose the circuitry was considerably simpler (Figure 2-6) consisting of a battery switch, a resistor output connector, and an OCP71 phototransistor. According to Robert Pickard (Zoology Department, University College, Cardiff, Wales, personal communication, 1975), phototransistors can be used by connecting the emitter to the negative terminal via a 10 kΩ resistor, connecting the collector to the positive terminal, and then recording across the resistor as shown in Figure 2-6. He further advises that the voltage used be 9 V: this voltage may then be changed and the 10 kΩ resistor may be altered to pick up the desired output voltage. This circuit was originally used to measure dragonfly nymph ventilation where the light beam was interrupted by abdominal movements of the restrained nymph.

Klostermeyer and Gerber (1969) also employed a photoelectric device to measure activity. Their device was two Clarex CL603 photocells mounted to provide two light paths in a nest. The resultant record on two

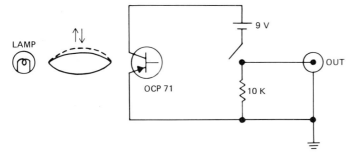

Figure 2-6. Circuit used by Pickard and Mill (1974) to record the respiratory movements of dragonfly nymphs.

channels of a Rustrak® event recorder could distinguish entry from exit and determine how long the bee was in or out of the nest. From a detailed knowledge of the behavior of the bee, *Megachile rotundata,* a nest was constructed of wood and Lucite® containing the two photocell light paths. The separate behavior patterns during nest preparation by the bee were provisioning, ovipositioning and capping.

The device built by Klostermeyer and Gerber enabled precise times during nesting behavior to be measured. The influence of external factors such as temperature and light intensity was also measured and evaluated. An estimate was made of the amount of alfalfa seed set by female leaf-cutters in experiments in and around Prosser, Washington.

Since conducting these studies of photocell recording techniques, Klostermeyer and his colleagues changed to an electronic recording balance technique to obtain a continuous record of weight changes in nests of *Megachile rotundata* during nesting behavior (Klostermeyer et al., 1973). The newer technique improves upon the photocell method by providing more information but is slightly more complex. Since the new method also uses a photoelectric detector, it will be described here.

A drinking straw 4.0, 4.8, or 6.2 mm in diameter, and cut to 10 cm in length, served as the bee nest. This was supported horizontally on one end of a beam. The opposite end of the beam contained a short soft-iron pin extending vertically into an air core solenoid. Silicone oil in a dash pot served to clamp the movements of the beam. As the weight of the nest changed, the beam tipped on its center fulcrum, and this was recorded by a light beam moving across two photocells.

Movement of the light beam was recorded as an error voltage that was amplified 30,000 times. The current of the air core solenoid was proportional to the photocell error voltage and this current caused an attraction of the soft-iron pin which, in turn, rebalanced the beam to a null position. These position corrections were recorded potentiometrically on a recorder calibrated in milligrams.

Full scale of the available recorder was set at 100 mg, and, since the fully loaded nest weighed over 1 g, an automatic tare was devised. A limit switch on the recorder pen activated a motorized device which dropped glass beads into a taring pan attached to the solenoid end of the beam balance. This reset the recorder periodically.

The electronic recording balance was used outdoors. High winds caused an unsteady trace, but the record was still interpretable. The equipment could be used for other cavity nesting insects.

Results using this method showed a close association between the weight of provisions and the weight and sex of the progeny. More females (fertilized eggs) were produced in larger holes, and fertilized eggs were placed in the inner cells and males in the outer cells in a single nest. More provisions were placed in cells with female eggs.

Although the article by Klostermeyer et al. (1973) offered to provide details of instrumentation, two requests by the author for circuit diagrams were not successful. Klostermeyer (personal communication, June 1976) has, in fact, discouraged any attempts to secure more details by taking the position that photocell circuitry is readily available, and that the more complicated weighing device could be built around any analytical electrobalance. He further felt that it would be too much trouble to examine the devices as finally evolved, since numerous changes had altered the original circuitry.

Spangler (1969) also developed a photoelectric device to detect hive entrance activity of honey bees. The design used RCA photocells. Several days were allowed for the bees to adjust to the modification of the entrance (Figure 2-7). The counting circuitry reset immediately to count a subsequent bee. On rare occasions, a bee would block the light path, back up, then start forward again to be counted twice.

Kerfoot (1966) reported a relatively simple circuit for detecting bee activity that was used in the field with a solitary halictid bee, *Agapostemon texanua*. The detecting device was an LDR-C1 cadmium sulfide cell in series with a 10 kΩ variable resistor, a Sigma relay, 11F 1000G Sil, and 27 V DC.

A portable ant counter was developed around the photocell method for recording the passage of leaf-cutting ants, *Acromyrmex octospinosus* Reich and *Alta cephalotes* L., in Trinidad (Dibley and Lewis, 1972). Capacitance actographs were considered to be too delicate for field use. Instead, a photocell method was adapted from similar techniques used by others for counting ants in the laboratory.

The final ant counter operated outdoors in the wet tropics, used minimal current from 12 V DC sources and could be adjusted to count ants or, when aimed high, to count leaves carried by ants. The current was limited by resistances in the lamp circuit and by the transistor arrangement. With no ants present, the photocells were illuminated biasing the first stage to

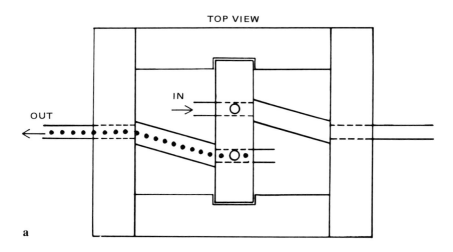

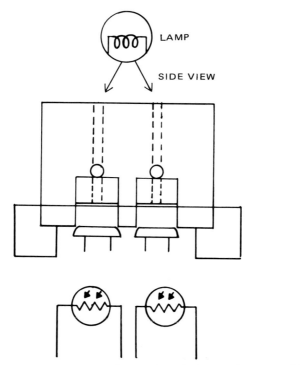

Figure 2-7. Bee entrance block with light path and photoresistors used for determining the passage of honey bees into and out of a hive (Spangler, 1969). (a) Top view, dots show exit pathway; two circles show the position of the light path perpendicular to the entry and exit pathways. (b) Side view shows the light source above and photoresistors below the entrance block.

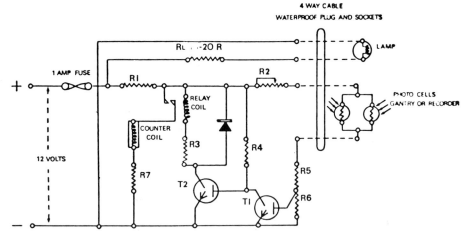

Figure 2-8. Circuitry for the detection of the passage of ants in the field (Dibley and Lewis, 1972). R1, 50kΩ optional; R2, 50kΩ; R3, 500kΩ selected for relay; R4, 10kΩ; R5, 1kΩ; R6, 270kΩ; R7, 20kΩ selected for counter; T1 and T2, OC 139; Relay, Radiospares miniature open type 15; Diodes, 1SJ50; Counter, 12-V DC; photocells, Photain Controls CdS Type 5 SP5-4; Recorder, ORP 60 Mullard; Lamps, 6-V 33-A MES Tub 11 mms.

conduct (Figure 2-8). This in turn biased the second stage to off and deenergized the relay. Thus no current flowed in the counter circuit when the light beam was unobstructed.

A helpful section was provided by Dibley and Lewis covering counting efficiency and sources of error. Maximum efficiency was attained only when slowly moving ants carried a leaf of only one sort. The device could not distinguish two ants simultaneously interrupting the light beam and counted one individual more than once for various reasons. Most importantly, calibration was accomplished by comparing counted results with direct visual observation.

A discussion of sources of error in the Dibley and Lewis paper is highly instructive. It helps provide an idea of the types of problems encountered in using a purely static measuring device and emphasizes how variable certain insects are under certain conditions. The experimenter is often presented with behavior of some complexity in the field. The most bothersome in connection with light beams has to do with edge effects. If the ant stops in the light beam just at threshold for counting, for example, the circuit design will determine whether a multitude of counts occur or just one. We encountered this problem in another context (Miller et al., 1971) and had to redesign the photocell circuitry so that successive counts required the photocell to be completely covered then completely uncovered.

In the latter example, houseflies were held over a Delrin® wheel (Fig-

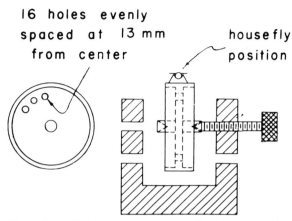

Figure 2-9. Physical arrangement of a wheel machined from Delrin® which acts as a substrate for a walking house fly. Sixteen holes drilled in the wheel interrupted a light beam as the fly moved the wheel. After Miller et al. (1971).

ure 2-9). Movement was recorded as the number of times the Delrin wheel interrupted a light beam. The circuitry (Figure 2-10) included transistors Q_1 and Q_2 arranged as a Schmitt trigger circuit. Variation of light input to the photoresistor caused a corresponding variation of input voltage to the gate of Q_1. With each zero crossing of this voltage the trigger changed state. Each change of state impressed a positive pulse upon the gate of Q_3 via the RC networks and diodes leading from the drains of Q_1 and Q_2. Transistors Q_3 and Q_4 comprised a monostable multivibrator, which generated a positive output pulse of about 0.5 sec duration for each

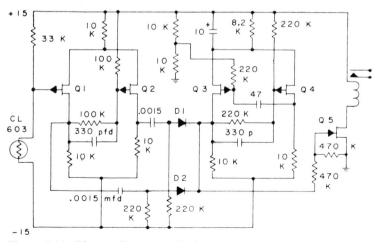

Figure 2-10. Photocell actograph circuitry accompanying the fly wheel shown in Figure 2-9. After Miller et al. (1971).

input pulse received. The output pulse, impressed upon the gate of Q_5, drove the latter into conduction and thereby actuated the relay in its drain circuit. Closure of this relay operated a printer-counter. This complicated circuit was necessary since the printer-counter could accept only a brief contact closure for each event.

Smith and Campbell (1975) also reported a photoelectric device of a slightly more advanced nature. Both directional information and size discrimination were possible by positioning two light beams. The device was used successfully to record movements of nocturnally migrating cave crickets in Spider Cave at the Carlsbad Caverns National Park (New Mexico).

A red light photocell was used by Cooper et al. (1971) in studying the movements of ants, *Camponotus herculeanus,* confined to a Petri dish area. Each dish containing seven ants was placed over two photocells. One photocell placed peripherally or near the edge of the dish recorded the "radial" activity and a second photocell positioned centrally recorded activity at the center of the dish.

It is not clear if the ants were able to detect this red light. However, circadian rhythms were not being measured; only response to an exposure to drugs was measured. Data were collected for 19 hr intervals in each test. Although no statistics on the significance of variations in data were presented, the authors reported that each of the following—water, mescaline, pentobarbital, perphenagine, and amphetamine—produced a unique and separately identifiable qualitative pattern of activity while the pattern of activity from tranyleypromine was indistinguishable from the activity of ants exposed to water.

An important point to be learned from the Cooper study was that in all cases the "radial" activity measured at the edge of the circular dish exceeded that recorded near the center of the dish. In the control study $63 \pm 3\%$ of activity was radial. This information would be useful for photocell placement when studying other insects confined to dishes. The output of the photocell trigger amplifiers (Figure 2-11) was used to drive a cumula-

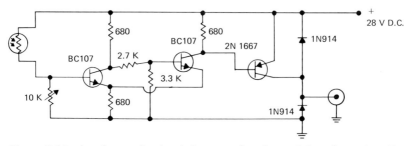

Figure 2-11. An electronic circuit for counting the activity of ants in a Petri dish by the interruption of a light beam (Cooper et al, 1971). The trigger point is adjusted by the variable resistor at the base input.

tive recorder and results were presented as the number of photocell traverses occurring over time.

A combination of the rocking-box actograph and photocell methods was described by Dreisig (1971) who constructed a small cage balanced on a pivot with a small dark paper attached to one end. As the cockroach, *Ectobius lapponicus,* moved about the cage, the dark paper interrupted an infrared light beam directed toward a photocell which in turn alternately energized a relay and then tripped the pen of an event recorder.

The Bailey Instrument Co. (C.P.B. Proffitt, Chief Development Engineer) has suggested two photocell circuits for use with the Bailey LA-1 amplifier for photodetectors. The LA-1 is an all-purpose amplifier and is relatively inexpensive (U.S. $145.00 in 1975). Photo diodes may be connected directly to the input terminals of the LA-1 (photovoltaic mode) and the output terminals of the LA-1 are connected to a voltmeter or a relay which energizes at a particular light level. A suggested relay for the output circuit is the Calectro D1-960, a 1.5 V 10 mW relay with contacts rated at 0.25 A, 12 V DC/36 V AC. The photovoltaic mode produces a logarithmic voltage response to light intensity.

When used in the photoconductive mode, a linear response of voltage to light intensity is obtained (Figure 2-12). The photodiode shown in Figure 2-12 is the Centronic OSD50 from Bailey, or an equivalent obtained from major sources (UDT, Texas Instruments, Hewlett–Packard, General Electric, EG&G, etc.).

Another somewhat less accurate possibility for measuring light intensity with the LA-1 includes a less expensive cadmium sulfide photo resistor (Figure 2-13).

There are innumerable other possible photocell circuits available in kit form from electronics stores. One example of this is kits manufactured by EICO (283 Malta Street, Brooklyn, New York 11207). These are relatively inexpensive and can be used for setting up experiments for

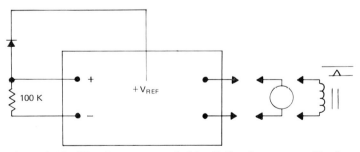

Figure 2-12. Circuit recommended by Bailey Instrument Co. for use with a photo diode and the Bailey LA-1 amplifier for a light beam actograph. The photo diode is the Centronic OSD50 or equivalent. A choice of two outputs is shown, voltmeter or relay.

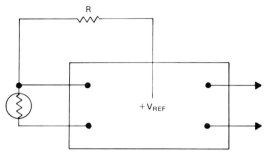

Figure 2-13. Circuit recommended by Bailey Instrument Co. for use with a photoresistor in light beam actographs.

preliminary evaluation since the kits are easily and rapidly assembled. Part IV, 2 describes optical devices used with transducers. Much of the same technology applies to actographs or transducers.

4. Acoustical Actographs

Measuring insect activity by the sounds they emit or by the noise they make has been accomplished independently in several cases. Sound-proofing to eliminate extraneous sounds is the major drawback to this method.

a. Microphonics

A rather elaborate description for recording mosquito flight sound was given by Jones (1964). The method used a 2.5 cm crystal microphone and the Tektronix 122 amplifier. However, any amplifier could be used for this with a gain of around 1000.

Noise levels were measured at up to a peak of 15 μV from the microphone inside the specially designed acoustical container. Signals from flying mosquitoes were recorded at 40 to 400 μV at the microphone output. Since *Anopheles gambiae* females produced a fundamental frequency of 440 Hz during flight and had a prominent first harmonic at double this frequency, filters can be set for the harmonic between 800 and 1000 Hz and a considerable reduction in noise level can be attained as a result.

Jones has recently improved this early design (Jones et al., 1967) and is continuing to improve the circuitry even further (M.D.R. Jones, School of Biological Sciences, University of Sussex, Falmer, Brighton, Sussex, BNI 90G, U.K.) (Jones and Reiter, 1975; Jones et al., 1972). The latest version of the Jones flight actograph circuit (Figure 2-14) uses a loud

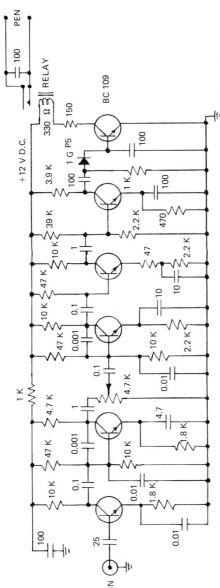

Figure 2-14. Microphone actograph of M. D. R. Jones. The circuitry includes a preamplifier of two stages including a sensitivity control, followed by four amplifier stages ending in a switching circuit which drives an event pen. All capacitors are in mF unless otherwise noted. The microphone is a 2.5-in. loudspeaker of 35 or 75 Ω or an ordinary 200-Ω electrodynamic telephone microphone element. The relay is a 330-Ω Radio Spares® relay operated by 6 V DC.

speaker element rather than a microphone as the input transducer. For the output circuit, any relay operating on 6 V and 20 nA would do, and was used to switch 12 V and about 0.5 A. The capacitor across the relay contacts prevented sparking which would interfere with operations.

This same method has been used to record noises made by house flies landing on paper cups and to record their buzzing sounds during convulsions due to insecticide poisoning (Miller, 1976). For this a Midland No. 22-106 lapel microphone was used. Eight-ounce Dixie cups (No. 2168-SE) containing house flies were placed on top of the microphones. The cups were placed on plastic foam used in shipping electronic instruments and an ordinary 2-liter glass battery jar was inverted over the cups to shield out extraneous noise. The microphones were connected to Isleworth A-101 amplifiers with gain set at 100 and filters set at 200 Hz low cutoff and 5 kHz high cutoff. Sounds were tape recorded at 1-7/8 in. (Tanberg 1000 tape recorder). After an hour or more, collected data were played back into an Ortec time histogram analyzer (Models 4620 and 4621) at 7.5 ips. The time histograms were then sampled from the Ortec stored data and recorded directly on Brush 220 pen recorders.

Since some tape recorders have various drive speeds, it is possible to collect data at 15/16 ips recording speeds and to play back collected data at 15 ips for a time condensation of 16:1. In other words, data that were collected in an hour could be played back in less than 4 min into other devices. Alternatively, sounds produced by insects in the kHz range could be tape recorded at high speeds and played back at low tape drive speeds. This would enable high-frequency signals to be recorded on ordinary pen recorders with frequency responses in the low hundreds or below.

Thus, while microphone recording of insect sounds suffers from extraneous noise, it can be done with relatively inexpensive equipment and, other than the recording arena, requires no special construction.

Another example of an acoustical chamber constructed for insect actographs is that by Pesson and Ozer (1968) and Pesson and Girish (1968). Here a tube containing a granary weevil in a bean was glued to a microphone. The microphone was hung by its lead wire to rest inside an acoustically insulated chamber.

Another vibration technique was used by Pesson et al. (1971) to measure the activity of house flies during poisoning by various insecticides. Five to 15-day-old adult female house flies of a susceptible strain were placed in clear plastic tubes 4 cm in diameter and 3 cm high. An electrical acoustic device was attached to one face of the tube. Vibrations were amplified and recorded with sufficient sensitivity to render the projection and retraction of limbs. This and the preceding device were derived from the work of Andrieu (1968) who reviewed a number of actograph techniques and developed several himself.

Another type of vibration actograph was developed by King (1973) to measure the disturbance caused by landings of the cocoa mirid, *Distantiella theobroma*. Microphones of the dynamic type, not crystal or ceramic, were obtained from Henry's Radio, Ltd. (303 Edgeware Road, London, England). Four dynamic microphones were dismantled by removing the diaphragms, and nonferrous pins were soldered to the armatures. The modified microphones were then glued to foam rubber pads and mounted to an aluminum frame so that the nonferrous pins projected toward a cage with aluminum gauze sides. The pins were waxed to the aluminum gauzes so that any collisions of the flying insects with the gauze caused a movement of the microphone element via the nonferrous pin.

Four microphones mounted as above were connected in parallel and their signals were amplified, operated a Schmitt trigger circuit, then were amplified again to drive the pen of a milliammeter recorder operated at 7.6 cm/hr (Figure 2-15). The maximum sensitivity of the recorder was found to depend very much upon how the microphones were set up, i.e., where the armature was placed in relation to the permanent magnet. The most sensitive position had to be found by trial and error and the attachment of the pin to the armature and cage side had to be very carefully done so that there was no resultant tension on the pin (King, personal communication).

A light-dependent resistor (LDR, Figure 2-15) was used in a unique manner in the output circuitry. This device was mounted in the experimental chamber, and it enabled a DC potential to be recorded which was proportional to light intensity. The momentary pulses recorded from cocoa mirid acitivity were recorded on top of this information.

b. Vibration

Another example of recording movement of insects by the vibration method was that developed by Leppla and Spangler (1971) for recording

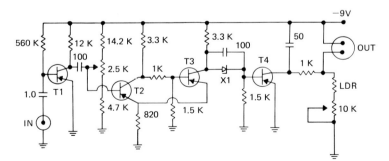

Figure 2-15. Circuitry for a microphone flight actograph for the cocoa mirid. T1, 2, and 3 are AC107 and T4 is OC72. X1 is a 4.3-V Zener. LDR: cadmium selenide light dependent resistance. All capacitors in mF. After King (1973).

activity of Pink Bollworm moths, *Pectinophora gossypiella*. This system was relatively simple compared to the method of King described above.

The moth cage was 8 cm in diameter and 8 cm high and was made of galvanized steel ends with a bronze mesh screen for the cylindrical sides. The cage with an access hole drilled in one end and eight hooks soldered to the top and bottom was suspended at the center of a 44× 44-cm wood frame by rubber bands.

The cage was suspended on all sided, i.e., pulling up from four corners at the top and stretching from four points on the bottom to each of the eight corners of the wood support frame. Suspension by rubber or other elastic bands appears to be the preferred method of isolating a cage of insects from substrate vibrations. King's method also used elastic supports to isolate the cage.

Pink Bollworm activity was detected by a Pixie® silica chip transducer (Endevco Corp.). An equivalent silicone strain gauge is manufactured by Pye Dynamics Ltd. (Park Avenue, Bushey, Herts. England). The silicone chip, a piezoresistive device, is delicate and must be handled with care. Leppla and Spangler (1971) used a straightforward circuit connection (Figure 2-16).

An ordinary terminal lug weighing 1.25 g was soldered to the top end of a vertically positioned Pixie strain gauge. The bottom end of the Pixie was soldered to a support which was rigidly attached to the side of the metal moth cage. In this case, the support holding the Pixie was a phono

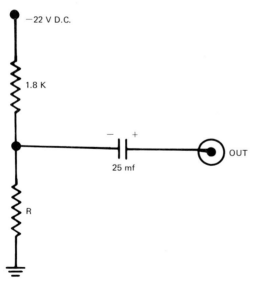

Figure 2-16. Circuit elements used to connect the Pixie® silicone semiconductor strain gauge to recording instruments which monitored the landings by Pink Bollworm moths in a flight cage (Leppla and Spangler, 1971).

jack mounted on a 29 ×29-cm tin can. The can was screwed to the moth cage and the phono jack provided electrical connection to the transducer.

Signals developed by changing resistance of the Pixie strain gauge (Figure 2-16, R) upon movement of the insects. Disturbances caused a resonance of the cage and the plastic support. Vibration of the silicone strain gauge depended somewhat upon the weight of the soldered terminal lug attached to the top of the strain gauge. Leppla and Spangler (1971) reported two resonances near 150 and 500 Hz.

The signal developed across the 25 μF capacitor was fed through a preamplifier to a General Radio 1521-B graphic level recorder modified for long period usage with a ballpoint pen cartridge. Chart speed was set at 6.35 cm/hr. Since this recorder required 100 mV of peak signal for full-scale deflection, the transducer signals were amplified with a gain of 40. Although the actograph could be operated from 22.5 V DC dry cell batteries, a line-operated power supply was used instead.

Rather than constructing a preamplifier above for amplifying signals from the Pixie strain gauge, an inexpensive unit is available from ordinary stereo shops which is designed to preamplify low-level signals from magnetic cartridge pen recorders for amplification by units designed for ceramic cartridge phonograph cartridges. Typical examples from local stores around Riverside, California, are the Realistic Model 42-2930 preamplifier sold by Radio Shack, and the Herald Model AM-50A preamplifier (Herald Electronics, Lincolnwood, Illinois 60645) sold by Pacific Stereo stores. The latter lists an input of 10 μV, an output of 0.5 V, a band width of 30 to 20,000 kHz, and a signal-to-noise ratio of 60 db. The advantage of these devices is their low cost [U.S. $10.95 (1976)].

The Pixie transducer has also been used to measure stress forces (Sykes et al., 1970) where two silicone transducers were connected in the arms of a bridge circuit to give additive effects. The output of the bridge was amplified by a type 1221 microvoltmeter (Comark Electronics, Rustington, Sussex, U.K.). Forces down to 0.5 mg were measured with the output linear to 10 g.

Another device with a vibration principle was developed by W. Stendel (Bayer AG, Institute for Chemotherapy, 56 Wuppertal-1, Postfach 13 01 05, West Germany) for use in recording scraping sounds produced by ticks, *Boophilus microplus,* as they move on filter paper. The movements or activity were recorded during poisoning treatments to determine the type of action by various acaricides. Further information concerning this technique is not available at present. Dr. Stendel is revising the instrumentation and has written that he will publish detailed information when the new version is performing well (W. Stendel, personal communication, February 1976). The entire device was mounted inside a lead chamber to reduce interference.

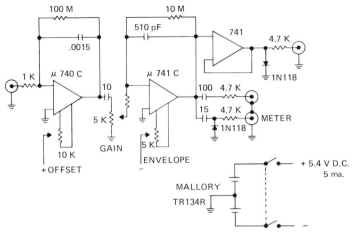

Figure 2-17. An unpublished circuit by R. Pence used for measuring the vibrations from an isolated abdomen of the worker honey bee (cf. Pence et al., 1975).

Mr. Roy Pence developed a device to record the vibration movements of an isolated bee abdomen, *Apis mellifera*. The pickup device was a piezoelectric cartridge and the amplifier circuitry is extremely useful as a general device. Since there is no description of the device in the publications associated with this work (Pence et al., 1963; Pence et al., 1975), it will be described here (Figure 2-17).

Using the Pence vibration-sensitive device, two output signals were recorded. The first (Figure 2-18a) included the total vibration signal. The second part of the output signal (Figure 2-18b) was the envelope signal or only the positive deflections of the basic signal. Either form of the signal could be useful in operating counting instruments which contain amplitude discriminator circuits. No elaborate shielding or soundproofing was used with this device.

Another piezoelectric device was reported by Medioni (1964) who developed a circuit to measure the activity of *Drosophila melanogaster*. The cage was constructed of cigarette papers and transparent plastic whose total weight did not exceed 75 mg. This rested on a piezoelectric device (Pathé Mélodyne 53) which was electrically connected to six stages of amplification to drive a counter circuit. Although Pence did not use shielding for the bee abdomen measurement, Medioni did resort to insulation to exclude outside vibrations.

Many insects with chewing mouth parts are not amenable to electrical monitoring of the type described in Section 6c. Instead Marcos Kogan (University of Illinois, Natural History Survey, Urbana, Illinois 61801) developed a recording apparatus (Figure 2-19) based on vibrations produced during chewing (Kogan, 1972, 1973).

A test leaf was trimmed to a 2 cm strip and stretched on a mylar

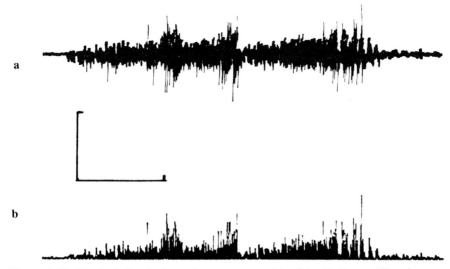

Figure 2-18. Typical signals from the output a and b of the R. Pence vibration actograph shown in Figure 2-17. (a) shows the total signal and (b) is the envelope or monophasic portion of (a). Calibration: vertical, 200 mV; horizontal, 5 sec.

membrane. The base of the leaf and the petiole was pinched between the plexiglas support stand and a hinged plate which was held down by a rubber band (the rubber band is not shown in Figure 2-19). The tip of the leaf was clipped with a heavy paper clamp (not shown in Figure 2-19) which pulled the leaf over the mylar roller. The leaf petiole was kept moist in a reservoir between two layers of cellulose sponge. A square plastic lid 55 x 42 x 14 mm was placed over the leaf and contained the test insect. Notches kept the lid from touching the leaf.

An insect pin was glued to the sensor probe and projected through a hole in the mylar membrane to rest on the bottom surface of the leaf. A flat piece of aluminum foil of 1 cm diameter was glued to the tip of the pin to ensure a greater area of contact between the leaf and the sensor. The sensor was a Model DB "Polysonic Detector and Bionic Sensor" (C.W. Dickey Associates, 705 W. Nittany Avenue, State College, Pennsylvania 16801).

The specimen holder was mounted on a 5-cm-thick pad and the band pass filter of the detector was set to pass 200 to 5000 Hz. No other acoustical shielding was necessary. The detector was connected to a Heathkit Model EV B20 chart recorder.

Leaves used in the original study were obtained from soybean plants of the variety "Clark C3" (Kogan, 1973). The chewing of the larvae of five insects was recorded successfully: fourth stage larvae of Mexican bean beetle, *Epilachna varivestis* and a noctuid, *Autographa precationis*.

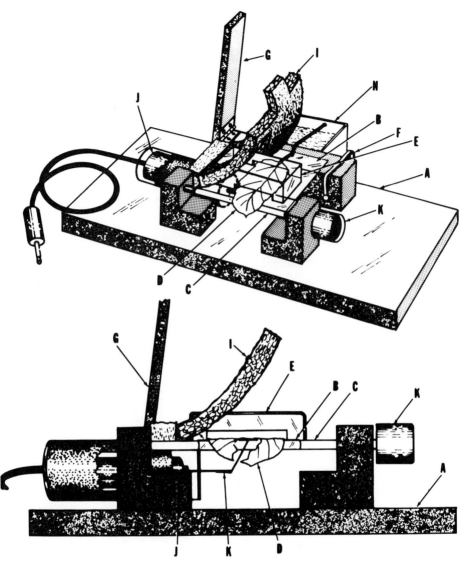

Figure 2-19. Physical appearance of the device used by Kogan (1973) to record masticatory disturbances from chewing insects; a feeding actograph. A, Plexiglas base; B, mylar membrane; C, plastic rod; D, test leaf strip; E, half box enclosure of test insect; F, rubber band fastener of the hinged plate; G, hinged plate in open position; H, moist reservoir partially filled with wet cellulose sponge; I, cellulose sponge pad; J, "Bionic Sensor" enclosed in foam-padded plastic cylinder; K, knob that permits control of tension rod, C; L, sensor probe with bent insect pin.

Locomotory movements were recorded as larger potentials than chewing, but the chewing vibrations were characteristic patterns.

Generalized circuitry for use with either vibration or audio actographs has been developed by E. W. Hamilton (USDA Laboratory, Post Office

Box 16545, Gainesville, Florida 32604) (Figure 2-20). An input amplifier (Figure 2-20a) accepts signals from a microphone, a speaker, or a photograph cartridge. Choice of output circuits include a low-power relay to give contact closure with each pulse of activity (Figure 2-20b), outputs to drive an AC-powered device (Figure 2-20c and d), or a pulse taken from

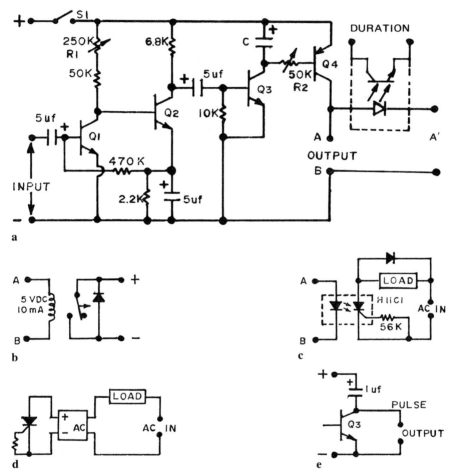

Figure 2-20. Generalized circuitry devised by E. W. Hamilton adaptable for microphonic, vibration, or audio actographic devices (Hamilton, 1976). Overall circuit amplifier including an output (A′) for duration counting is shown in (a). (b) through (e) show various output options. The diode in (b) is connected across a load containing inductance to prevent false counting. (c) provides half-wave AC to the load, and circuit (d) provides full wave AC to the load. In (d) AC shows full-wave bridge rectifier; Q1, 2, 3, 2N2484 or ECG 123A; Q4, ECG 159. In (b) an Allied electronics relay No. AMP 2009 was used. In (c) the H11C1 was later changed to H11C2.

Q_3 to drive a solid-state electronic counter. Output c provides a half-wave AC voltage, and output 'd must be used when a load requires full wave (a synchronous clock motor for example). The Hamilton circuit provides an optoelectronic coupling to drive a separate timer circuit which could record the summated duration of activity (Figure 2-20a, output A^1).

c. Ultrasonics

Ultrasonics have been used successfully to record insect activity. Alton Higgins produced insect activity devices a few years ago under the company name Alton Electronics, and some laboratories still have his equipment (Holloway and Smith, 1975). Unfortunately, it is rather difficult to contact the person responsible for the rights to the Alton Electronics motion detector. According to Dr. Norman Leppla (at the USDA, Post Office Box 14565, Gainesville, Florida 32604), Mr. Higgins may still be contacted at 4430 Creighton Boulevard, Pensacola, Florida 32504, and he will provide custom-made electronics devices upon order; however, my own attempts to contact Mr. Higgins have been unsuccessful to date. The address of Alton Electronics referenced in Finger (1972) is an old one, not useful.

The advantage of ultrasonic detection is the complete absence of light and lack of restraint conditions, which can interfere with behavior. Thus strict control over all environmental conditions is possible. Alternatively, no measurements of the possible effects of ultra sound on insect behavior are known, except for the more famous perception of bat sonar signals by noctuid moths in flight.

For ultrasonic detection of the activity of cornstalk borers, *Elasmopalpus lignosellus* (Zeller), the insects were placed in 2.54 cm diameter acetate tubes (Holloway and Smith, 1975). The ends of the tubes were closed with the probe and detector of an Alton Electronics motion detector. A 2% honey–water solution was provided from a saturated wick in the side of the tube. Mating had a pronounced effect on male locomotion activity with early scotophase activity unaltered and late scotophase activity inhibited.

An ultrasonic actograph was developed independently by Peter B. Buchan (ARC Unit, Zoology Department, Cambridge University, Cambridge, England) and was reported in a recent study (Treherne and Willmer, 1975). The following description was taken largely from correspondence with Mr. Buchan who has kindly given permission to reproduce these materials.

Ultrasonics fall in the frequency ranges around 40kHz. The wavelength of the sound wave is about 7.5 mm and consequently the minimum size of the insect to be detected is limited. *Dicheirotrichus pubescens*, which are between 5 and 6 mm in length, have had their activity monitored with success (Treherne and Foster, 1977).

The detection principle depends upon a standing wave or pressure pattern set up within the confines of an arena. This pattern is disturbed by any movement within the box and is seen by the passive transducer as a variation in amplitude. The standing wave pattern is not uniform partly as a result of the geometry of the arena, and a greater disturbance may be caused in one part of the box compared to another part for the same activity or movement.

Of the two arenas tested the one shown in Figure 2-21a is the most straightforward and has been used to monitor *Periplaneta americana,*

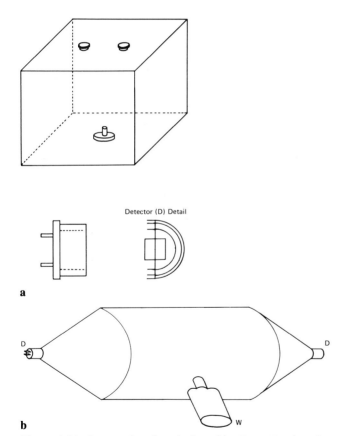

Figure 2-21. Insect chamber designed by Peter Buchan for use with ultrasonic actographs. (a) Cockroach arena. Overall dimensions are 22 cm long, 18 cm high, 20 cm wide. A water dish is placed in the middle of the floor with paired transducers overhead. (b) Container for measuring house fly activity. The central area is a 15-cm-diameter Plexiglas cylinder 15 cm long and capped by 60° funnels with detectors, D, mounted in the spouts. W, water source. P. Buchan, 1975, personal communication.

Musca domestica, and also the pupae of *Musca domestica.* Also shown is some detail of the transducer construction. The system in Figure 2-21b shows an attempt to make the pattern more homogeneous and was designed specifically for monitoring *M. domestica.* The solid angle of radiation is claimed to have a value of 60° by the supplier, hence the utilization of the 60° funnels as a means of confining the beam.

A later design using a parabolic reflector with the transducers fixed at the focal point was also tried and found successful but no extended program was carried out. The parabolic reflectors were in fact the old design of reflectors used in automobile headlamps.

The amplitude variation caused by any movement in the containers is amplified and detected. To avoid vibration and low-frequency interference from external sources, the tuned amplifier is by far the best choice (Figure 2-22). Sending a phase signal from the 40kHz source to a phase-sensitive detector would appear to be an excellent circuit for the purpose. In practice, however, it proves to be ultrasensitive, and drifts continually as a result of small changes in the relative positions of the transducers caused by ambient temperature variation.

After detection the remaining signal is available for recording directly or may be further processed and arranged in a final form suitable for analysis. Two methods of dealing with the information have been used. The first detects the signal and shows when the activity has reached or exceeded a predetermined level. The number of and distance between the

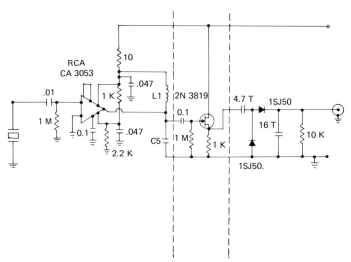

Figure 2-22. Circuitry used with ultrasonic actographs designed by Peter Buchan. A block diagram shown in the inset indicates position. L1 was selected for resonance with C5 which was about 1000 pF. 4.7 T and 16 T are tantalum capacitors. All capacitors are in mF unless otherwise shown.

pulses may be noted for future analysis (Comparators, Burr Brown, 1973. *Applications of Operational Amplifiers, Third Generation Techniques,* pp. 109–114). The second method turned the signal into a pulse form where the frequency of the pulses is proportional to the signal amplitude. These pulses are easily integrated and the integrator reset at desired intervals (*Precise Rectifiers,* absolute value circuits, Burr Brown, pp. 120–125. *Operational Amplifers, Design and Application.* Voltage to Frequency Convertors, Burr Brown, 1971, p. 408).

Specifications for the Buchan ultrasonic transducers are listed in Table 2-1. The transducers were purchased from Harry's Radio (303 Edgeware Road, London W2 1BW, England).

5. Electrostatic Methods

Electrostatic methods have been developed for recording insect activity in a few cases. Edwards (1960) caged five male and five female *Calliphora vicina* blowflies in a cubic steel box 9 in. (22.86 cm) on a side. A conducting probe was hung in the box and connected to the input of a Kiethly Model 600 electrometer. Activity was measured as deflections on a 100 mV Varian Model G-10 pen recorder. Records from this arrangement showed drift, and the measurement was probably sensitive to humidity contributing to drift. It is not clear why Edwards did not use a blocking capacitor to eliminate slow potential changes and record only fast signals coincident with flight activity. The electrometer actograph has been extremely versatile for Edwards (1964) who was able to adopt it to a wide variety of recording situations with relatively simple electrodes.

A later electrostatic detector was constructed by J.A. Miller et al.

Table 2-1. Ultrasonic Transducers Specification[a]

Series resonant frequency	36 kHz
Impedance at resonance (measuring voltage 3 V)	300 Ω
Receiving sensitivity at resonance ($R_{in} = 10\ \Omega$)	1 $\mu A/\mu bar$
Parallel resonant frequency	39 kHz
Impedance at resonance (measuring voltage 3 V)	15 kΩ
Receiving sensitivity at resonance ($R_{in} = 1$ MΩ)	4 mV/μbar
Nominal capacity at 1 kHz	2200 p^F

[a]Transducers available from: Henry's Radio, 303 Edgeware Road, London W2 1BW, England; price £6.00 per pair.

(1969). A screened room $3 \times 3 \times 24$m housed the setup with temperature and humidity controlled. The electrode was a 1 to 3 cm O.D. copper tube 23 cm long and rubber-jacketed into a brass-screened flight cage. A $10^9\Omega$ resistor shunted the electrode to ground to prevent charge accumulation. After amplification by an electrometer, all deflections greater than 20 μv from the base line were counted as flights. Horn flies *Haematobia irritans* or Stable flies, *Stomoxys calcitrans*, were placed in the flight cage for activity recording.

Movement of flies toward the electrode caused positive voltage deflections and movements away from the electrode caused negative voltage deflections. The magnitude of recorded voltages depended on the static charge on the fly, the speed of flight and the proximity of the flight to the electrode.

It was found that a cage made of brass screen gave flies a net positive charge. When the walls were lined inside with glass, flies leaving the glass surface contained a net negative charge and recorded voltages were opposite those described above, such that leaving the glass surface and approaching the electrode caused a negative voltage deflection as recorded by the electrometer. Use of glass also reduced the amplitude of responses. There was no attempt to determine if the electrostatic recording procedure was detected by the flies tested or whether it affected their behavior.

6. Electronic Methods

a. Capacitance

The principles of capacitance devices are discussed in the early works by Schechter et al. (1963) and Löfqvist and Stenram (1965) and were predated by the abbreviated report of Backlund and Ekeroot (1950). All subsequent electronic methods are essentially similar to the early versions in that the insect is placed in a position to alter the inductance or capacitance of a tuned resonant circuit. Movement would then detune the resonant circuit and the shift of frequency is detected and amplified to ultimately drive a recording device.

There are two disadvantages to these methods as reported in the literature. One is that they are somewhat involved electronically and tuned circuits are difficult to deal with for the novice. The second problem is that stray capacitance usually requires shielding of some sort, although this is rarely limiting.

The advantages of capacitance methods are similar to those of ultrasonic methods including a lack of restraint, no lights, and the fact that the test insect does not have to come in contact with electrodes or move something, although the measuring chamber represents somewhat of a spacial constraint. In addition, capacitance methods are extremely sensitive with outside interference easily minimized. Therefore electronic actographs are well suited for insects confined to a small space or in situations in which insects are made to encounter the recording electrodes by moving about in a larger space.

The utility of capacitance methods is emphasized by the number of times such devices have been developed for measuring insect movement (Backlund and Ekeroot, 1950; Schechter et al., 1963; Grobbelaar et al., 1967; Löfqvist and Stenram, 1965; Luff and Molyneux, 1970; Stange and Hardeland, 1970; Hardeland and Stange, 1971, 1973; Evans, 1972; Hilliard and Butz, 1969; Fondacaro and Butz, 1970; Dutky et al., 1963). An unusual combination of biological interest and electronics knowledge at a fairly sophistocated level is required if capacitance devices are to be constructed.

Apart from applications specifically designed for use with insects, capacitance devices have been manufactured for some years to measure the movements of mammals (Ögren, 1970). *"Animex"* is produced by LKB (LKB Instruments, 12221 Parklawn Drive, Rockville, Maryland 20852) and *"Varimex"* is produced by Columbus Instruments (950 N. Hague Ave., Columbus, Ohio 43204).

Recently Columbus Instruments developed *"Entomex,"* a device to record insect movements which can detect movements of nymphs of the cockroach *Symploce capitata*. Various cage designs allow *Entomex* to be adapted for measuring other insect movements, and Columbus Instruments has advertised that it is willing to design recording chambers around specific applications.

The capacitance actograph described by Dutky et al. (1963) has undergone some modification with each application. However, the device, a thermocap relay, has been reportedly used by Fondacaro and Butz (1970) to record circadian rhythms of *Tenebrio molitor*, and by Hilliard and Butz (1969) to record circadian rhythms of the American cockroach *Periplaneta americana*.

The Thomas Register lists Niagra Electron Laboratories (1-11 Rochambeau Avenue, Andover, New York 14806) as the manufacturer of the thermocap relay. Using maximum sensitivity settings, the Thermocap is said to be sensitive to the movements of the antennae of the cockroach (A. Butz, Biological Sciences, University of Cincinnati, Cincinnati, Ohio 45221, personal communication, May 1976).

Mr. Jim Gram from Niagra Electronics (July 1976, personal communication) related that the Thermocap Relay® is now being redesigned as Model 400 which will be a solid-state device with production planned for early fall 1976. The new model will have a remote probe for use at a distance from the chassis.

The Thermocap was designed primarily as a temperature detector, but has the unusual property of being sensitive to insect movement. The new version of the relay will provide an output of a TRIAC solid-state switch rated at 120 V and 10 A. No circuit connection is provided which would give an output voltage proportional to capacitance changes, but the circuitry is amenable for such an output according to the designers.

Grobbelaar et al. (1967) reported a circuit designed to be versatile, sensitive, inexpensive and drift free. It reportedly was used to monitor the activity of hamsters, aquatic insects, and small insects such as *Drosophila* and thrips (Figure 2-23).

Some specialized construction was necessary to make the coils for the tuned circuit. L1 was 24 turns of 24 gauge enamel-covered wire on a 1.27 cm cylindrical frame with a piece of ferrite rod 7.62 cm long used as the core of the coil. L2 was 55 turns of 36 gauge enamel-covered wire on a 1.27 cm diameter cylinder tapped 15 turns from one end (B and E on Figure 2-23). A powdered iron core for L2 was used for fine tuning. L3 was 31 turns of 24 gauge enamel-covered wire wound and the same 1.27 cm diameter frame as before with a ferrite rod similar to that of L1 used for tuning.

Construction of the movement-sensing electrodes is straightforward as long as certain precautions are kept in mind. A pair of parallel plates does not give a large variation in capacitance for a given movement. For best results the electric field between the recording electrodes should be made nonuniform. Grobbelaar et al. (1967) made one electrode a wire spiral wrapped about 16 times around a 50 ml nonconducting container. The second electrode was 12 turns of wire fixed inside the container and about 3 mm away from the inner surface of the container. The exact geometry of the electrode wires depended upon the size of the insect being studied, but the area between the electrode wires was the most sensitive.

Evans (1972) modified the capacitance actograph of Grobbelaar et al. (1967) for recording the activity of Tenebrionid beetles, *Alphitobius piceus*. A special cage was designed containing the sensing electrodes on two parallel opposed surfaces. The top surface was 11.5 × 11.5 cm square and a wire was laid back and forth across the top surface covering a circular area of 10.5 cm in diameter. The second electrode, in order to establish a nonuniform electric field, was wound as a spiral starting at the center and ending at the edge of the bottom surface in a circle also of 10.5 cm in diameter.

Evans found that the closer the wire coils or loops were wound or the closer the top and bottom surfaces were held together, the more sensitive the capacitance detector was to small movements. The thickness of the spacer separating the electrode surfaces was determined by the size of the insect being studied. The entire activity cage was housed in a temperature- and humidity-controlled chamber.

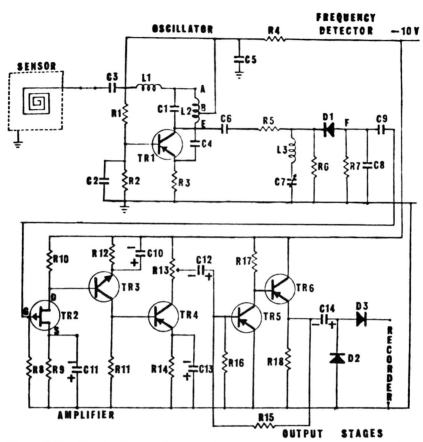

Figure 2-23. Circuit diagram for capacitance actograph published by Grobbelaar et al. (1967). The circuit values are: C1, 10 pF; C2, 0.1 mF; C3, 250 pF; C4, 20 pF; C5, 0.1 mF; C6, 56 pF; C7, 0 to 10 pF trimmer; C8, 0.25 mF; C9, 3 mF 50-V paper; C10, 500 mF 30 V; C11, 100 mF 30 V; C12, 40 mF 500 V used for low leakage; C13, 500 mF 30 V; C14, 1000 mF 30 V; R1, 3.9 kΩ; R2, 1.8 kΩ; R3, 1.5 kΩ; R4, 1kΩ; R5, 27 kΩ; R6, 150 kΩ; R7, 100 kΩ; R8, 470 kΩ; R9, 6.8 kΩ; R10, 10 kΩ; R11, 5.6 kΩ; R12, 3.9 kΩ; R13, 5-kΩ potentiometer; R14, 4.7 kΩ; R15, 100 kΩ; R16, 22 kΩ; R17, 2.2 kΩ; R18, 1 kΩ. D1, D2, D3, all OA85. TR1, OC44; TR2, 2N3820; TR3, OC140; TR4, OC44; TR5, OC71; TR6, OC71.

The Grobbelaar circuit was modified to be operated by an AC power source instead of batteries for prolonged experiments. Evans amplified the output of the capacitance actograph to drive a comparator which switched a relay when a selected voltage signal was exceeded. The relay contact closures were counted and printed at hourly intervals on a Sodeco counter-printer.

A variation of the circuitry of Grobbelaar et al. (1967) was published by Stange and Hardeland (1970) for measuring the activity of various

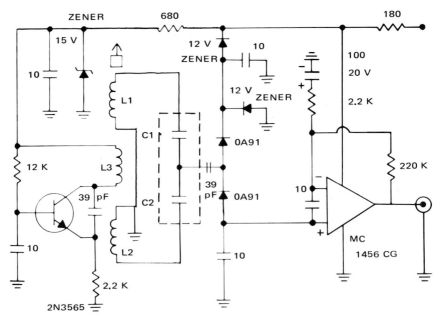

Figure 2-24. An improved version of the Stange and Hardeland (1970) capacitance actograph. L1, L2, and L3 are coils wrapped as described in the text. C1 and C2 are wires mounted in the recording chamber (Figure 2-25). D3 and D4 are 12-V Zener diodes and D5 is a 15-V Zener diode. (Stange, 1976, personal communication.)

species of *Drosophila* including *D. melanogaster* and *D. azteca*. This, in turn, was modeled after a device developed by Machan and Himstedt (1970) for recording the locomotion activity of newts and salamanders.

The basic difference in the latter two circuits is a long-term stability and decrease in outside interference which was accomplished by connecting the detector capacitors in a bridge circuit. Stange and Hardeland (1970) reported sensitivity and reliability in operations covering several months with low maintenance.

The circuit bears a superficial resemblance to the one of Grobbelaar et al. (1967), but the major difference is in the oscillator stage. The Stange circuit has been modified and the newer version is shown in Figure 2-24. The measuring chamber or box was constructed of 1-mm-thick "Astralon" with inner dimensions of 1.5 × 40 × 17 mm. Three parallel 0.35-mm diameter insulated copper wires were placed on one of the flat sides of the box in grooves and connected together as common plates of capacitors C1 and C2 (Figure 2-25).

Two additional wires were placed on the opposite flat side in the spaces between the first three wires and these acted as the outside plates of C1 and C2, respectively. These five wires then formed capacitors C1 and C2

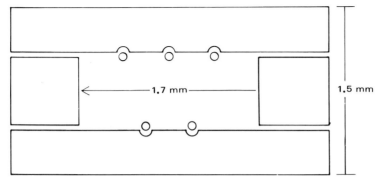

Figure 2-25. A cross-sectional view of the chamber used with the Stange and Hardeland (1970) capacitance actograph described in Figure 2-24. Approximately to scale, the dimensions of the inner part of the chamber are given in the text. (Stange, 1976, personal communication.)

and, together with coils L1 and L2, formed the measuring bridge of the capacitance-sensitive circuit.

The bridge activates itself with a frequency of about 30 MHz allowing detection of the extremely small capacitance changes due to insect movements. It is also necessary to adjust L1 and L2 to a slightly off-balance position of the capacitance bridge for maximum sensitivity. L1, L2, and L3 were formed on an 8-mm diameter coil form with a VHF core. An 0.35 mm diameter insulated copper wire was wound 20 times each for L1 and L2, and 8 winds for L3 with each of the coils about 3 mm apart. L3 was wound in between L1 and L2 on the form with the middle of L1 and L2 grounded.

The figure description for the original circuit diagram in the Stange and Hardeland (1970, abb. 2, p. 402) article is mislabeled in the portion dealing with construction of the coils, so that L1, L2, and L3 should be read L3, L1, and L2, respectively (Stange, 1976, personal communication).

One of the more recent and complex capacitance actograph circuits was reported by Luff and Molyneux (1970). The specific advantage of this method is a lack of specially constructed coils for tuned circuits. Capacitance detection is based on two points (A and B of Figure 2-26) having the same capacitance to a common ground point (C of Figure 2-26).

Nine-centimeter diameter plastic Petri dishes served as the activity chamber. Three detector wires were taped to the outside of the dish on the bottom (Figure 2-27). Their distance apart was gauged roughly as the length of the insect under study.

In the capacitance detector, Al oscillates at about 1600 Hz as a Wein bridge oscillator stabilized by D1 and D2. Al supplies one arm of a

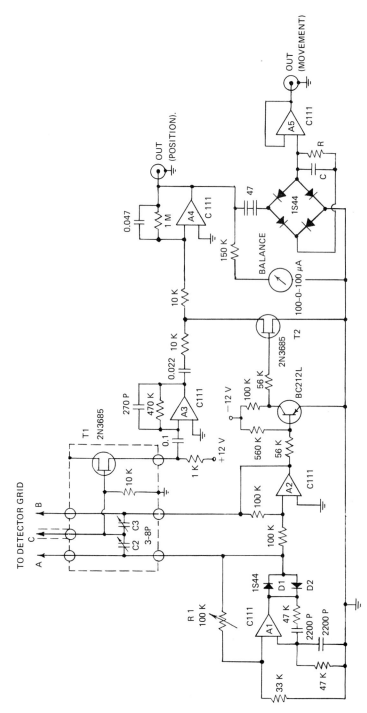

Figure 2-26. Revised circuit version of the Luff and Molyneux (1970) capacitance actograph (Luff, 1974, personal communication).

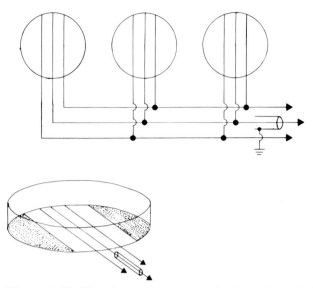

Figure 2-27. Chamber arrangement of electrodes of the Luff and Molyneux capacitance actograph shown in Figure 2-26.

capacitor bridge. The other arm is supplied in antiphase from A2 which phase inverts the original signal with unity gain. The transconductance of T1 is not controlled by negative feedback and the sensitivity of the device depends upon the transconductance of T1.

T2 acts as a phase-sensitive detector where the mean value of voltage pulses at the source terminal is a measure of bridge unbalance. A4 amplifies and averages these voltage pulses and the feedback loop of A4 offers a choice of three sensitivities from 1 to 3 Hz.

To set into operation: adjust R1 to obtain 10 V peak-to-peak at A1 output; close S1, connect the detectors to the activity chamber, and adjust trimmers C3 and C4 until the output from A4 is zero; then open S1 and select an appropriate frequency response range (S2). In practice, greater movements in the activity chamber cause greater voltage output. The field effect transistor is important in that its transconductance value determines sensitivity. The lead to the gate of T1 should be kept short and shielded.

A "position" record is obtained from the output of A4 which is further rectified and applied to A5 via an RC filter. The output of A5 is "movement" record. Selection of an RC time constant of 10 sec gave meaningful signals from Carabid beetles, *Pterostichus modidus* (F.).

The adaption of the Luff–Molyneux detector for movements on irregular surfaces of smaller or larger insect than carabids should be attainable. However, its great advantage is that measurements in complex environments are practical since the detector would not be unbalanced by local variations in the moisture content of the air. Thus movements in saturated air should be possible to detect because of the unusually good stability of the detector circuit.

b. Inductance

Running wheel actographs (described in Part II, 2, a) copied from designs used for vertebrates have been adopted for recording activity of crickets and cockroaches (Sokolove, 1975; Roberts, 1960). However, the stick insect, *Carausius morosus,* proved able to move around the wheel container without registering counts (Godden, 1973).

Instead a magnet 1/8 in. diameter and 1/2 in. long was waxed to the dorsum of the *Carausius.* Twelve hundred turns of 28 copper wire wound on a 2 in. diameter core 7 in. long provided the cage space containing the insect and a bit of food plant.

Movements induced currents in the copper wire which were amplified and recorded on an Esterline Angus pen recorder. Any ordinary AC amplifier would presumably work for this application as long as it could detect low frequencies since *Carausius* movements are ordinarily slow and deliberate during feeding.

c. Resistance

Measuring the activity of mosquitoes presents a challenge to the experimenter. Of the actographs described above, the rocking-box method was considered too delicate and impractical for large numbers of mosquito adults. Sound recordings similar to those developed by Jones (1964) were also considered impractical for large numbers; in addition sound recording would not detect walking activity only flight.

Powell et al. (1966) developed a simple resistance circuit for recording mosquito activity which could be expanded for use on a large scale in a genetic selection program. The measuring chamber was a Plexiglas surface 12 × 12 cm upon which inverted glass funnels of 8 cm diameter were placed.

Bare silver wires of 100mm diameter were arranged as parallel grid lines across the floor of the chamber with 1 mm spacing. Alternate wires were attached in series with a resistor of 10 mΩ, a Dynograph recorder, and a 1.5 V DC dry cell battery (Figure 2-28). The Dynograph recorder was listed as a Beckman Offner Type R Dynograph (Beckman Instruments Co., 2500-TR Harbor Boulevard, Fullerton, California 92634). This is a Beckman physiological recorder which is produced by their Offner division, but presumably any general laboratory recorder with

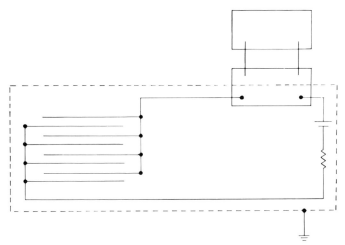

Figure 2-28. The circuit arrangement used by Powell et al., (1966) to detect the activity of mosquitoes by intermittent closure of a simple electric circuit.

high-impedance input and high gain would do. The short circuit current was 1.36×10^{-7} A, but recorded changes were of the order of 10^{-9}A.

When the inverted funnel was kept clean, mosquitoes were unable to land on the glass; therefore, flights were infrequent and the adults spent most of their time on the measuring grid. Humidity was increased by drilling 50 holes of 1 mm diameter in the chamber floor and placing the chamber over a Petri dish of water. Electrical interference was reduced by placing a Faraday cage around the entire measurement chamber.

Ball (1972) developed and used an electrical contact scheme for recording movements of *Diabrotica virgifera* adults and nymphs of *Oncopeltus fasciatus* and *Periplaneta americana*.   He found that *P. americana* was not suitable for this device, possibly because of cuticular waxes insulating the electrical contacts. Cages were $6.5 \times 6.5 \times 2.5$ cm and food was provided in the cages in vials.

Electrodes were arranged in the cages so that test insects contacted both electrodes at once. This depended in part upon the size of the insects used, but symmetrically interspersed strips of copper printed circuit sheet provided convenient contacts for larger insects. (Figure 2-29a, inset).

The circuit diagram redrawn from Ball (1972) (Figure 2-29) includes only one channel; however, the number of channels and therefore grids used could be increased by connecting each new grid to the power supply and B7 and by providing a separate amplifier for each grid. Thus, for each cage, a new amplifier is needed, but the power supply can be used

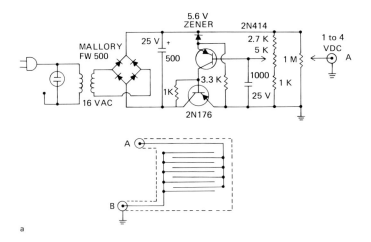

a

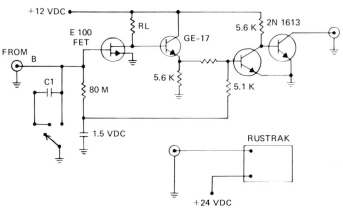

Figure 2-29. Electric actograph circuit used by Ball (1972). The inset in (a) shows the arrangement of bare wire electrodes and the power supply to drive the electrodes. (b) shows the amplifier and output circuit suitable to connection to a solenoid recorder. C1 is selected for the insect, 100 pF, 30 V to start for *Oncopeltus fasciatus*. The value of R_L is also selected for the insect used.

for several cages. All leads to the grid electrodes had to be thoroughly shielded to reduce electrical interference.

The capacitor C1 was selected for each insect used. For adults of the western corn rootworm, *Diabrotica virgifera,* C1 was removed. The value of RL depended on the characteristics of the field effect transistor, E-100. Briefly, RL was adjusted until the drain voltage was +2 V DC and the gate voltage was −0.2 V DC. This adjustment matched the circuit to the voltage supplied by the regulated power supply. RL usually fell between 10 and 22kΩ. The output of the amplifier was taken to a Rustrak (Gulton Industries, Gulton Industrial Park, East Greenwich, Rhode

Island 02818) four-channel event recorder (Model 92). The Rustrak recorder was connected as shown in the inset in Figure 2-29b. When activated by a contact across the measuring grids, the final stage transistor, 2N1613, conducts and the Rustrak pen solenoid is effectively connected to 24 V which energizes the pen to produce a count. Otherwise the amplifier represents a high resistance to the recorder pen circuits.

Ball (1976, personal communication) considered this electrical contact actograph to hold no particular advantage over capacitance actographs; however, each recording method has its own peculiar advantages for a particular insect. As reported in the paper (Ball, 1972), the contact method was not successful with nymphs of *Periplaneta americana*, but did function satisfactorily for *Diabrotica virgifera* adults or for nymphs of the large milkweed bug, *Oncopeltus fasciatus*. (1) Feeding Actographs The electrical method of measuring drinking or sucking is essentially the same principle that was used by Goedden (1973) to monitor water supply to the food source for the stick insect, *Carausius morosus*. Feeding actographs using electronic or electrical circuits have been developed for a number of different insects (Lipton and Sutherland, 1970b; Kashin 1966; Fredman and Steinhardt, 1973; Getting, 1971; Kogan, 1973; Smith and Friend, 1970; Kogan and Goedden, 1971).

The simplest type of feeding apparatus is equivalent to the mechanical actograph. Lipton and Sutherland (1970b) developed a "trophometer" consisting of a plastic collar, food, and a washer contact (Figure 2-30). Food, F, is attached to a 28 gauge steel wire, S, 10 cm long, connected as a pendulum. Any disturbance of the food pellet causes electrical contact between the steel wire and washer, W, which is counted on an event re-

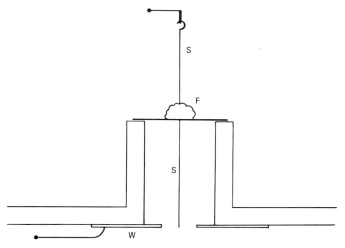

Figure 2-30. Trophometer feeding actograph using electrical contact to detect activity (Lipton and Sutherland, 1970b).

corder (Esterline Angus, Model A620X). Since electrical contact causes a spark, the light interference caused by the spark may be reduced by baffles and by painting the surfaces black.

Lipton and Sutherland (1970b) determined that a pendulum provided with an innocuous weight instead of a food pellet also recorded activity. They were able to distinguish between searching activity and feeding activity since the latter caused periods of pen activations lasting 5 min or more while random disturbance or nonfeeding activity caused only brief periods of even recordings. The trophometer was used with adult male cockroaches, *Periplaneta americana*.

P. Kashin (1966; Kashin and Wakeley, 1965; Kashin and Arneson, 1969) devleoped a method to record the mosquito bite. Each of the parts of feeding behavior including probing, penetration, salivation, engorgement, and withdrawal could be separately distinguished. The method was developed as a test for chemicals with repellent or attractant properties to *Aedes aegypti*. Engorgement on anesthetized mice was used previously as a test score; however, this yes-no observation provided no information on sampling behavior by the mosquito. Salivation is thought to be the mechanism by which mosquito-borne diseases are passed and therefore salivation is rather important as a bioassay criterion.

Electrical contact could easily be distinguished in the chart records as an excursion from flat baseline recordings. Engorgement was observed to coincide with low-frequency and low-amplitude peaks in the record, suggesting the functioning of the cibarial and pharyngeal pumps acting in concern with the posterior and anterior pharyngeal valves.

While no overt behavior was observed to accompany salivation, Kashin tentatively assumed that a pattern of peaks oppositely directed compared to those seen during engorgement were the signals accompanying salivation (Kashin, 1966). Furthermore, Kashin interpreted large excursions in his recordings and prolonged feeding times (up to 10 min) as pool feeding or feeding from a small hemorrhage. Small excursions and comparatively short feeding periods of about 3 min were interpreted as clean engorgement directly from a capillary in the host.

The method employs 50 or 100 mesh bronze screen as one side of a mosquito cage; the screen is connected through 16 MΩ to a 1.35 V DC Mallory mercury battery (RM-1-R or RM-12-R) to one input of a Sanborn Model 320 recorder (input impedance: 0.5 MΩ). The second terminal of the recorder is attached to a hypodermic needle inserted into the tail of an anesthetized mouse. The bronze screen rests on the unshaved mouse, and the mosquitoes complete the recording circuit by standing on the screen and probing the mouse.

One mosquito commonly caused 3 mV deflections during feeding. From circuit considerations, this represented a 6.0×10^{-9} A current flowing through the mosquito during feeding. There was no indication that such a small current affected the mosquito.

In an update of Kashin's original mosquito bitometer, two parameters were measured to determine a measure of repellency. P is the percentage of actual biting time and E is the percentage of mosquitoes engorged so that the sum of $P + E$ is the repellency index. P was measured as the total time biting divided by the total time of the test times 100%. In practice P proved difficult to measure so an automated device was constructed to help eliminate errors arising from manual counting (Kashin and Arneson, 1969).

A circuit developed for monitoring the feeding of *Rhodnius prolixus* was essentially the same as that developed for mosquito feeding with one important difference (Smith and Friend, 1970). A 5 mil (127 mm diameter) platinum wire was waxed through the cuticle and into the hemolymph of a fifth instar *Rhodnius* through a small hole punched in the dorsal part of the thorax. After allowing 7 days for recovery, the *Rhodnius* were placed on a piece of No. 60 mesh brass cloth and allowed to probe through a rubber diaphragm into a diet medium (Figure 2-31). Other diaphragms have been used for mosquito feeding devices including chick skin preparation and bovine intestine preparation (Badruche membrane) (Jones and Potter, 1972; Wills et al., 1974). These give variable success depending on the species of mosquito.

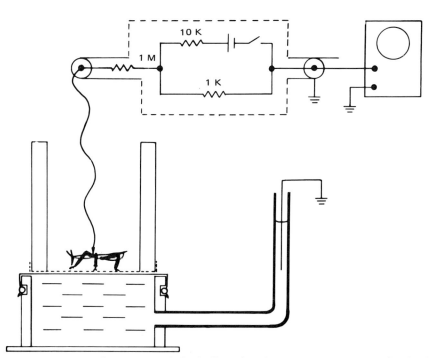

Figure 2-31. Feeding actograph including chamber arrangement and circuit elements for *Rhodnius prolixus*. After Smith and Friend (1970).

Air was eliminated from the diet chamber by filling liquid diet through the side arm, sealing the diaphragm cover in place , and then tipping to allow any trapped air to escape out the side arm. A similar diaphragm has been used to enclose defined aphid diets for aphid probing (*Myzus persicae*). Originally parafilm-M was used, but this was replaced by moist filter paper or Saran® wrap or thin silicone rubber when it was learned that the parafilm introduced phenolic and other substances into the aphid diets (A. Green, Zoology Department, University of Glasgow, Scotland, personal communication, 1973).

During feeding, the *Rhodnius* nymph completed a circuit as a variable resistance. Feeding responses were recorded as variations in voltage at the oscilloscope or these voltages were amplified and operated pen recorders. Because of the high impedance of the recording circuit, electrical interference was a problem and careful screening was necessary.

The *Rhodnius* contributed at least 1 MΩ resistance to the circuitry so that at most a current of 5×10^{-8} A flowed through the nymph. While this current caused no obvious effect on feeding, currents at 10 times this value did result in abnormal behavior. The resistors shown were chosen to produce the greatest signal for feeding with normal behavior. The voltage recordings obtained during feeding were calibrated by replacing the insect 1 MΩ and greater resistors with resultant voltage deflection noted.

For comparison between the diet chamber and host animals, the *Rhodnius* chamber could be placed on a rabbit with the diet electrode replaced by a hypodermic needle inserted near the cage. The rabbit ear was a convenient choice for a feeding site.

The major difference between *Rhodnius* feeding and the Kashin mosquito technique described above is the use of a screen to make contact with the test insect. Smith and Friend (1970) recorded resistances between the wire in *Rhodnius* and the brass screen of the cage and found large variations with no overt movement. They also noted large and irregular fluctuations in resistance during recordings of *Rhodnius* feeding when electrodes were connected between the brass mesh and the diet. Since this method of recording produced records that showed little resemblance with observed behavior of feeding, it was considered to be artifactual, and would be difficult to interpret.

With electrodes connected between the animal and the diet, patterns were recorded which were correlated first with maxillary probing; a second distinctive phase was correlated with activity of the pharyngeal pump. Similar results were obtained when the diet chamber was replaced by a rabbit. Measurement of aphid feeding activity has been accomplished by different groups. Perhaps the more extensive references on aphid feeding are from McLean (McLean and Kinsey, 1964,

1965, 1967, 1968, 1969; McLean and Weigt, 1968).   Although the recording of aphid feeding from plant leaves is similar in principle to monitoring bloodsucking insect feeding, the approach used by McLean and co-workers is somewhat unique.

The early version (McLean and Kinsey, 1964) impressed part of a 6.3 V AC filament transformer voltage across a leaf. An aphid rested on a metal grid and probed the leaf. The grid was connected to a Heathkit stereo amplifier (Model AA-181) and the output of the amplifier was rectified and pen recorded. This was similar to the final versions (McLean and Weigt, 1968); however, the same problems of artifact accompanying *Rhodnius* grid recording also hold for aphid recording.

To overcome artifacts, a wire was attached to the aphids.   The 10.16 μm (0.0004 in.) diameter wire was of pure gold (Bar 760, Secon Wire Division, Secon Metals Corp., White Plains, New York 10600). The wire was attached with a conducting paint used for printed circuits, GC No. 21-1/Walsco No. 36, 1/2 Troy oz. (G.C. Electronics, Rockford, Illinois, 61100). For these manipulations, aphids were held by vacuum to avoid problems from use of CO_2 anesthesia. Aphids were placed on 85 mesh tungsten over a vacuum orifice (Buckbee-Mears Co., St. Paul 1, Minnesota, 55101 for tungsten mesh).

Most recent versions of the aphid feeding actograph used a modern version of the Heathkit stereo amplifier, Model AA14 (McLean and Weigt, 1968, Figure 2-32 a and b). A full-wave rectifier consisting of four diodes, 1N1614, was included between the amplifier and pen recorder. All leads had to be carefully shielded with coaxial cable to reduce interference. If

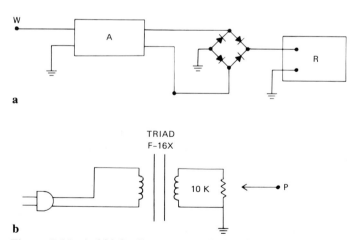

Figure 2-32. Aphid feeding actograph developed by McLean and Weigt (1968) using commercially available circuit elements. (a) shows amplifier, A, and recorder, R. A gold wire attached to the aphid was connected at W. (b) shows power source connected to the plant at P.

fluctuating line voltages were significant, a voltage regulator was used (Stabline, Model EMK 4105 or TRIAD K104).

Contact with the plant was made most recently by inserting a flattened 14-gauge copper wire (Figure 2-32 a and b) into the drain holes of a potted plant (K. S. McLean, personal communication, 1973). The plant, a broad bean, *Vicia faba*, was mounted against a Plexiglas board in such a way that one stem and leaf could be lightly clamped or held and an aphid could then be allowed to walk on the stem and probe the leaf. The size of the gold wire was selected so as to allow maximum freedom by the aphid. McLean's device was adapted to Lygus bugs or Tenebrionid beetles for which 38 gauge tinned copper was used as the electrical contact (Alpha Wire Corp.).

Measuring aphid feeding by use of transformer voltage applied across the aphid is one version of a general phenomenon. The same measurement has been developed using direct current power sources (Schaefers, 1966) and therefore would be more directly comparable to *Rhodnius* feeding and mosquito feeding. The use of DC eliminates most interference problems and also allows a more compact unit to be constructed.

A small rectangular silver-alloy strip was punched several times with a nail and then placed under a strawberry leaf. The leaf was pressed onto the ragged edges of the nail holes for electrical contact by means of two simple Plexiglas sheets, thus forming a sandwich.

A hole in one of the pieces of Plexiglas exposed the underside of the strawberry leaf. An aphid, strawberry aphids of the genus *Penta-trichopus*, was held by vacuum on a nylon mesh, attached to a gold wire (Secon Metals) of 7.6 μm diameter (0.0003 in.) by acetone-thinned silver print (G. C. Electronics) on the abdomen. Then the aphid was placed on the exposed leaf surface.

The aphid wire was attached to the positive terminal of a DC strip chart recorder (Bausch & Lomb, VOM 6). The negative terminal of the recorder was attached to the negative terminal of two series-connected 1.4 V mercury batteries. The positive terminal of the batteries was connected to the silver leaf electrode.

More recent refinements of this aphid feeding device have resulted in a circuit very similar to that used by Smith and Friend (1970). The only differences are use of a 1 or 10 MΩ variable voltage divider and a 100 MΩ resistor connected in parallel with the recording device. Either a Tektronix 564B storage oscilloscope or a Heathkit pH recording electrometer was used to record voltage changes during aphid feeding. (M. E. Montgomery, New York Agricultural Experimental Station, Geneva, New York 14456, Entomology Department, personal communication, 1974.

There are problems with DC recording since polarizing potentials can develop (McLean and Weigt, 1968). However, the basic principle appears to be widely adaptable. It has been used for aphids, *Rhodnius,* leafhoppers, *Empoasca fabae,* and the pear psylla, *Psylla pyricola.*

For insects with chewing mouthparts, the vibration technique described by Kogan (1973, cf. Part II, Section 4,b) was perfected, but an electrical resistance method was also developed (Kogan and Goedden, 1971). The latter was used to record chewing of chrysomelid beetle larvae, *Lema trilineata daturaphila* (Figure 2-33). The circuitry was similar to that used by Kashin (Kashin and Wakeley, 1965) and consisted of a copper screen floor (Dashed lines in Figure 2-33) and a 3.8 cm diameter tube for sides. A plastic disc (Figure 2-33, D) cut to fit inside the 3.8 cm diameter tube was mounted on a copper rod (Figure 2-33,C) and formed the roof of the chamber. The rod was held by an aligator clip 3.2 mm above the copper screen floor of the chamber.

A diet was held in place against the underside of the shield disc and a plastic spacer was mounted between the diet and the copper screen floor for support. The spacing of the chamber constrained the fourth instar beetle larvae to feed while in contact with the copper screen. Feeding completed an electrical circuit between the copper rod and the copper screen and this produced a variable resistance which was recorded as voltage fluctuations across a 2.5 MΩ resistor. A Heathkit Model EVA-20-26 servo recorder (Figure 2-33, R) set at 25 mV full scale was used to record voltage changes and the voltage source was picked off a 5 MΩ variable resistor which was connected in parallel with a 1.5 V battery.

A device to record both feeding and locomotion of *Lygus hesperus* was described recently (Sevacherian, 1975). The feeding device was essentially that of Kashin and Wakeley (1965) and consisted of a brass screen connected in series with a resistor, usually 1 to 5 MΩ, a battery, and then to one terminal of a 10-mV strip chart recorder (Heathkit EU-20v). The

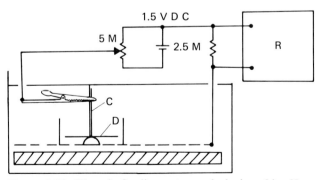

Figure 2-33. Electric feeding actograph designed by Kogan and Goedden (1971) including actograph arena.

other recorder terminal went to a plant leaf (cotton). With a nylon net between the brass screen and the cotton leaf, adult male or female *Lygus* probed the leaf to make contact and register voltage changes on the recorder.

The series high-resistance circuit did not appear to interfere with behavior of the *Lygus* bugs and no problems were noted with polarizing contacts. For long-term operation, the leaves lasted at least a week when kept moist and the *Lygus* continued to feed.

An elaborate drinking actograph was developed by Scheurer and Leuthold (1969) to monitor water intake by a cockroach, *Leucophaea maderae*. This consisted of two parts. A water-filled tube contained a series of platinum wire electrodes and a reference platinum electrode. When the water level fell below a certain level, a stepping motor was activated to turn a switch until the switch contacts again made electrical contact via the platinum wires in the water supply.

Not only water use was monitored in this manner; the presence of the cockroach was also registered since the water source was placed in such a manner that a light beam was interrupted during drinking. A photocell ordinarily illuminated by the light beam lead to a detecting and recording circuit to register visits to the water outlet.

7. Other Methods

a. Collection Methods

The collection of insect eggs oviposited over a period of time can be accomplished by a simple turntable device (Loher and Chandrashekaran, 1970) or by a sample changer adapted to receive eggs from caged insects (Godden, 1973). Quite elaborate mechanisms have been developed to collect eggs or record events over prolonged time intervals (Goryshin et al., 1973).

Most of the devices adopted for the collection of eggs or to measure the time of eclosion or hatching (Zimmerman et al., 1968) have been used in studies on circadian rhythms; however, a few biologists are interested in the time of oviposition or eclosion as information for its own sake which completes the detailed description of a species.

b. Temperature Methods

Flight actographs have been described above using the capacitance method (Grobbelaar et al., 1967), the electrostatic method (Edwards, 1960), and various acoustical methods including vibration (Leppla and Spangler, 1971), sound (Evans, 1972), and ultrasonics. Other devices to

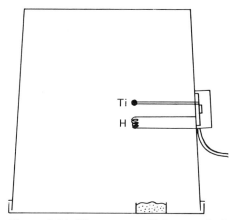

Figure 2-34. Transparent flight cage including food dish after Macauley (1972).

record flight involve the use of flight mills (Krogh and Weis-Fogh, 1952; Kishaba et al., 1967); however, these methods involve handling the insects and unnatural restraint (Macauley, 1972). Although the use of a wind tunnel to allow free flight has been adapted for studying aphid behavior (Kennedy and Booth, 1963), this method would be difficult to apply to other insects.

Several authors (Cavallin et al. 1972; Macauley, 1972; Hughes and Pitman, 1970) have described apparati based on the cooling effect of wing beats on small thermistors, either by disturbing the temperature gradient inside the cage or by cooling the thermistors already heated by the current flowing through it. An alternative method includes a separate source of heat consisting of a vertical helical coil (Figure 2-34, H) positioned below the thermistor (Figure 2-34Ti) to produce a column of warm rising air (Figure 2-34). When flight activity occurs, the column of warm rising air is deflected and the thermistor cools to the average temperature inside the cage and remains cool until stable air conditions allow the column of warm air to reestablish itself.

The thermistors used are the Bead type, I.T.T. type U 23US, Stock No. 1153E, with a diameter of 0.25 mm, connected through a simple amplifier with variable resistors in base and collector circuits to a chart recorder with full-scale deflection of 1 mA. Variations in ambient temperature cause irregularities in the baseline (Figure 2-35). Long-term variations could be reduced by using a bridge circuit employing another thermistor outside the cage (Figure 2-36, To). Some short-term variations, due perhaps to cycling of temperature-controlling equipment, can be reduced by placing a small block of metal in the cage to act as a heat storage system.

The warm air column is produced by winding 40 turns of Nichrome or resistance wire 32 S.W.G. around a Vitrosil glass tube 3 mm in diameter

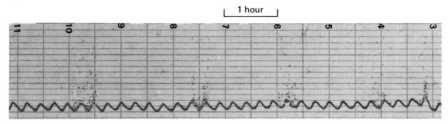

Figure 2-35. Flight activity of *Cydia nigracana,* the Pea moth. Time mark: 1 hr. Note the rhythmic fluctuations in room temperature occurring at a cycle of about 4 per hr and caused by the temperature-controlling devices in a constant temperature room. These traces were recorded from the Macauley flight actograph chamber shown in Figure 2-34 (Macauley, 1976, personal communication).

and by heated with about 1.2 V AC, preferably stabilized. This produces about 0.5 W of heat. The gap between the top of the heater and the thermistor bead is about 15 mm, although this can be varied to suit requirements.

As heat is being pumped continuously into the cage, there clearly needs to be some heat loss from the cage to avoid excessive temperatures inside it. The rate of heat loss will depend on the nature of the material from which the cage is made and the difference between the temperature inside and outside the cage. With use of inverted goldfish bowls made of clear acrylic plastic standing on a Plexiglas base, in an ambient temperature of 20°C, the temperature inside the bowls rises 2-3°C above this.

Using bowls with an average diameter of 19 cm and a height of 16 cm, it has been able to detect easily the flight activity of Pea moths, *Plusia gamma,* fruit flies, *Drosophila* species, and the smaller house flies, as well as hopping movements and wing fanning. The extent to which the air is disturbed, and hence the amount of signal obtained, varies with the distance of the flight from the column of warm air. There is also a time lag be-

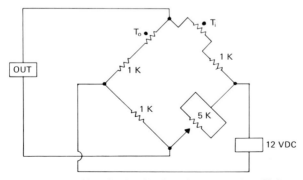

Figure 2-36. Circuit details for the Macauley flight actograph shown in Figure 2-34. The thermistors are Radiospares TH-B12. All resistors are 0.5 W.

tween the cessation of flight and the complete reestablishment of the air column. However this does not create much difficulty as it is possible to determine from the chart at what point the air column begins to reestablish itself. Durations of flight can, at suitable chart speeds, be measured to within 15 sec.

c. Kramer Sphere

Possibly one of the more sophisticated locomotory devices in recent times was developed by Dr. Ernst Kramer at the Max Planck Institut für Verhaltensphysiologie in Seewiesen, Germany (8131 Seewiesen) (Kramer, 1976). Refinements of this method were published recently by Erber (1975) (Figure 2-37) and Goetz (1972).

The insect used by Erber, *Tenebrio molitor,* is placed on a globe (earth globe available at stationery shops, 33-cm diameter). Four lamps (L) are positioned above the insect and centered around a position-sensitive detector produced by United Detector Technology (Pin SC 10). The lamps

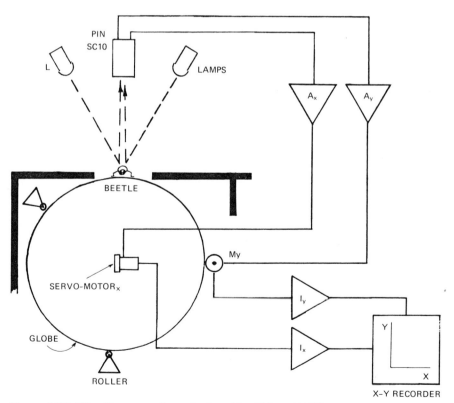

Figure 2-37. The Kramer sphere designed by Erber (1975).

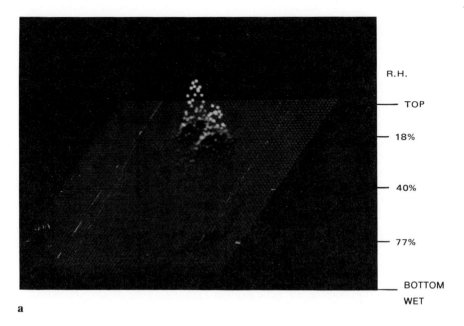

a

b

Figure 2-38. Four computer plots of radioactivity recorded by a dynacamera and depicting the position of four *Arenivaga investigata* cockroaches in a vertical column of sand 9 cm wide × 20 cm deep. A humidity gradient was maintained as shown. The frames were taken at time zero (a), then 30 min (b), 1 hr 50 min (c), and 6 hr 50 min (d) following introduction of the cockroaches into the container of sand.

c

d

Figure 2-38 (Continued).

were covered by RG 665 red filters and a white spot was marked on the thorax of the beetle.

When the insect wandered off-center, error signals in the x and y directions (Figure 2-37) were amplified (AX and Ay) and drove x and y motors (Mx and My) which rotated the globe until the insect was recentered and the error signal was zero. Whenever the globe motors were activated, the speed of rotation was integrated (Ix and Iy) and recorded on an $x-y$ recorder as the path of locomotion.

Thus while the insect remained in the same position on the globe, its path was accurately reproduced on the x-y recorder leaving the experimental arena free for manipulations.

The amplifiers Ax and Ay used operational amplifiers 1709, 1439, or 1741 for the input stage and ordinary power amplifiers for the output stages. The micromotors which drove the globe and microgenerators which provided the integration signals were obtained from Dr. Fritz Faulhaber, Schönaich, Württemberg, West Germany. The device of Dr. Erber required generators to be tailor made to fit the motors; however, Dr. Erber (personal communication, 1975) related that the Faulhaber firm now provides motors with gear and generators already attached.

It is necessary to select the motor to match the speed of the animal studied. The integrating amplifiers were also homemade, but employed OP-amps 1709, 1439, or 1741 as before.

Obviously this device requires some tailoring in construction and should be attempted only by those familiar with electronics and mechanics, the interested biologist would do well to seek a group approach or to enlist the assistance of an electronics technician.

Goetz (1972) (Max Planck Institut Für Biologische Kybernetic Tübingen, Germany) used a slightly different sensing system. *Drosophila* were attached to a sled which prevented flight and allowed walking. Any movement of the sled was immediately detected by a magnetic sensing device which sent error signals to servo motors. The motors turned the globe to reset the sled to the original position.

d. Dynacamera

Another relatively recent technique for measuring insect activity was pioneered by Dr. Eric Edney and his assistant Paul Franco for a highly specific application. Dr. Edney studied the behavior of a group of desert cockroaches, *Arenivaga* species (Edney, 1966; Edney et al., 1974). These cockroaches reportedly spend the day burrowed several feet beneath the surface of sandy areas of the Mojave desert in California. They dig to the surface at night to forage and feed, then burrow beneath the surface at day light.

The dynacamera which was developed for use with nuclear medicine proved useful for monitoring the position of radio-labeled *Arenivaga* in a container of sand where other methods would have been difficult to apply.

In an experiment which was conducted on November 20, 1972, four *Arenivaga investigata* cockroaches were placed in a vertical column of sand, 9 cm wide × 20 cm deep. The humidity gradient was 18% r.h. at 2 cm below the surface, 40% r.h. at the middle, and 77% r.h. at 18 cm from the surface.

Figure 2-38a, taken at time zero, indicates the cockroaches were near the surface. As they are close together, the cockroaches appear as just two peaks rather than four. After 30 min they were in the middle of the column (Figure 2-38b). By 1 hr 50 min all four cockroaches were easily distinguishable and were at or near the bottom of the gradient (Figure 2-38c). In frame 42, taken at 6 hr 50 min from the start of the experiment, little movement had occurred since the previous photograph (Figure 2-38d). The decrease in intensity of the peaks indicates the decay of the isotope, ^{99m}Tc with a half-life of 6 hr. Each photograph represents the counts accumulated over a 1 min period.

The humidities were determined with Ace electrolytic hygrometers (Yamato Scientific, Tokyo), each measured 25 × 12 × 3 mm and placed so as to impede the cockroaches as little as possible. The ^{99m}Tc was acquired as a salt in solution from a Molybdenum-99–Technitium-99m generator (New England Nuclear) and was applied to the dorsum of the abdomen as a flour paste (other methods of adhering the activity proved less reliable). The radioactivity was monitored with a Pickering IIC Dynacamera coupled to a Hewlett–Packard 2100A computer which compiled the counts and displayed the results on an oscilloscope from which the photographs were taken.

While the dynacamera is a highly specialized device, its use does suggest that the imaginative researcher has many possibilities for adopting modern technology to use in insect actographic recording.

Part III

Free-Moving and Tethered Preparations

1. Introduction

Perhaps the most direct method for measuring nervous activity in insects is capillary recording from chemosensory hairs. First introduced in the 1950s (Hodgson et al., 1955), this method consists of placing a small capillary tube containing an electrically conducting salt solution over a chemoreceptive hair, originally on the labellar hairs of Diptera, and recording with respect to a ground electrode in the head.

The second most direct measuring technique is that of recording the electrical activity of nerves or muscles by puncturing the cuticle in the strategic sites with sharpened metal electrodes or by inserting blunt or fragile wires through prepunctured holes in the cuticle. This latter method is the principal method used to record neuromuscular activity from intact insects. While microelectrode technology was described above as being difficult, extracellular recording as described here is considerably simpler and can be designed into laboratory exercises for students who have not dealt with electrophysiology before.

Measuring nerve and muscle activity from insects during the performance of normal behavior is a necessary step in establishing the neurophysiological basis of behavior. In the past this has been accomplished most often on dissected preparations where only part of the neuromusculature is intact. However, in several studies, insects have been left intact or nearly so while nerve and muscle activity has been recorded.

Perhaps the most readily accessible neuromuscular apparatus for measurement is that of flight. Adult insects can be rigidly held and flown either in wind tunnels or in static air, and flight behavior can be analyzed. This has been accomplished most notably for locust (Camhi, 1970; Wilson, 1968) but also for moths (Kammer, 1971; Obara, 1975), flies (Nachtigall and Wilson, 1967; Mulloney, 1969, 1970; Heide, 1975), and bees (Bastian, 1972).

A slightly more complex undertaking has been recording neuromuscular events which occur in intact insects during movement other than flight. Insect free-walking preparations have a rather recent origin.   Hoyle (1964) recorded muscle potentials extracellularly (electromyograms) from walking grasshoppers and he also recorded neuropilar potentials extracellularly (termed electroneuropilogram, ENG) by Hoyle in pioneering efforts described in Hoyle, 1970. Other work involving freely walking insects appeared shortly in which potentials were recorded from muscles involved in singing of crickets (Bentley and Kutsch, 1966), walking in cockroaches (Ewing and Manning, 1966; Pearson, 1973, 1976), and courtship in male grasshoppers (Elsner and Huber, 1969). Stout (1971) was able to record from cricket central connectives during free movement.

The technique of recording electrical activity from freely moving insects reached something of a plateau with the work of Howell Runion (Runion and Usherwood, 1966).   An energetic and imaginative man, Runion enjoys a familiarity with electronics which borders on genius. He builds his own amplifiers and other electronic equipment and has brought a characteristic sophistication to his studies in insect physiology whether in freely walking locusts (Runion and Usherwood, 1966; Usherwood and Runion, 1970), locust sensory recording (Runion and Usherwood, 1968), or endocrinology (Runion and Pipa, 1970).

Runion   developed   recording   methods   which   combined   electromyograms with electrical activity recorded extracellularly from axons which were wrapped by small diameter wires used as electrodes (Figure 3-1). This technique allowed the activity of inhibitory motor units to be recorded for the first time during locust walking. Myograms alone gave no hint as to the role of inhibitory units during walking (Ewing and Manning, 1966; Hoyle, 1970).   By recording nerve impulses in inhibitory axons during walking, it was determined that inhibitory units fired during the burst of excitatory nervous activity which suggested a role in modification of the tension caused by excitatory neuromuscular activity in locusts (Usherwood and Runion, 1970). By inference from partially dissected and   immobilized   preparations, Pearson (1973) found that the inhibitory units in the American cockroach were active late in the excitatory burst period. He interpreted this as inhibition being used to assist or hasten relaxation of particular motor units during repetitive events such as in rapid walking or running.

Some motor output activity in insects is strongly influenced by ascending sensory information; other motor activity is slightly affected or not

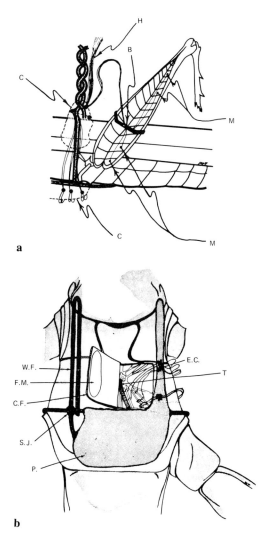

a

b

Figure 3-1. Drawings from Runion and Usherwood (1966) locust harness for free-walking preparation. (a) shows a side view with leg bracelet (B), wire electrodes for recording myograms (M), and harness lead wires (H); the dashed areas (C) show the position of cement for securing the main harness. (b) is a ventral view of the locust pterothorax with wire frame (W.F.) and harness solder joint (S.J.) cemented in place by adhesive (P.). A flap of cuticle (C.F.) is folded back to reveal the insertion scar of flight muscle (F.M.), a major tracheal branch (T), and electrode connections (E.C.) for recording extracellularly from major axon branches leaving the metathoracic ganglion.

affected at all by sensory information (Delcomyn, 1973). For this reason, the ability to record rather sophisticated nervous activity from insects which are intact and perfomring near normal functions is important.

Studies on the neurophysiological basis of behavior in insects have moved to a new level with the growing adaption of methods for recording from the somata of neurons in the central nervous system of insects. These methods necessarily employ glass microelectrodes and include not only recording of neuron activity but injection of dyes to later identify unequivocally the unit whose activity was recorded.

Numerous workers have exploited these methods to the point that the motor neurons from certain of the thoracic ganglia of *Periplaneta americana* and *Schistocerca gregaria* have been mapped and identified including the muscles innervated in the associated extremities (Hoyle and Burrows, 1973; Young, 1973).

The next level of study, that of interneurons and their role in the coordination of information, represents a further magnitude of complexity since the small size of many interneurons precludes recording from their cell bodies with present technology. Some interneurons are unusually large, such as cockroach giant axons (Parnas and Dagan, 1971) or the descending contralateral movement detector of locust (O'Shea et al., 1974). These are amenable to recording from various parts of the cell by microelectrodes.

Combining microelectrode recording from central neurons with an expression of some related motor behavior is an especially fruitful approach. This has already been attempted for ventilation rhythms in locusts with considerable success (Burrows, 1974, 1975). A preparation of the partially dissected American cockroach has been developed by Fred Delcomyn which promises to provide access to central nervous recording during movement of tethered insects (Delcomyn, 1976).

Because of nocturnal and secretive habits, the American cockroach is a rather difficult subject to deal with in free-walking preparation. In addition to struggling which is likely to occur with any restraint during handling, American cockroaches have a waxy cuticle layer and complex morphology which present certain technical problems to electrode emplacement. Derek Gammon (1977) used several implanted electrodes and a harness for monitoring central nervous activity in the American cockroach during poisoning. Delcomyn (1976) has reported central nervous activity from American cockroaches that were partially dissected and eviscerated, yet showed behavior indistinguishable from whole cockroaches. Since both of these studies are instructive, they will be described in some detail below.

2. Gammon Cockroach Preparation

The American cockroach preparation of Derek Gammon was developed during 1973–1976 as part of the research for the Ph.D. dissertation in the Zoology Department at the University of Cambridge. It was primarily intended as a preparation for studying the nervous system in an intact insect during insecticide poisoning.

a. Preparation of Electrodes

Lacquered copper wire (transformer wire) of 0.075 mm diameter was originally used for recording. However, electrode polarization became a problem when the same electrode was used for stimulation and recording.

Silver wire of 0.050 mm diameter (2 mil, 0.002 in.) was obtained from New Metals and Chemicals Ltd. to overcome the polarization problem. Insulated silver wire would have served better. After a thorough cleaning in petroleum ether to remove grease from the wire, bare silver wire described here was insulated by stringing 16 in. strands across a board and then painting with polyurethane varnish (FURNIGLAS P.U.-15, Furniglas Ltd.).

The cut ends of these insulated wires were chlorided by passing current for 30 sec then reversing polarity and repeating for a total of 3 min in 0.1 M HCl with 1.5 V DC. The silver chloride-plated wires were stored in 200 mM NaCl prior to use. If bubbles arose anywhere along the insulated portion of the wire during plating, the electrode was discarded.

The indifferent and stimulating electrodes were treated identically except that 1 to 2 mm of the electrode end was scraped before chloriding. Similarly the indifferent recording electrode was bared for 5 to 7 mm before chloriding.

Special stimulating electrodes were made for the cerci as the silver wires proved somewhat difficult to use due to limited longevity. Short (3 mm) sharpened tungsten pins were soldered (ordinary multi-core resin solder) to a length of small diameter copper wire (lacquered transformer wire, cleaned before soldering) and then the solder joint was covered with epoxy resin (araldite, a quick-setting version) for strength. Once inserted, this tungsten pin could be used for recording or stimulating and lasted for a period of some days. It was thought that activity recorded from the cercal pin represented afferent activity. Efferent impulses descending from the sixth abdominal ganglion were occasionally recorded when the pin was positioned near the base of a cercus.

b. Preparation of Cockroaches

Adult male cockroaches were kept overnight without food. Without using anesthesia, the cockroaches were pinned and stapled dorsal side down on a white rubber-based surface with care taken only to restrain and not to damage the cockroach. Struggling could not be avoided during electrode placement. To avoid permanent damage or irreversible changes in behavior, the operations were held to 2 hr or less.

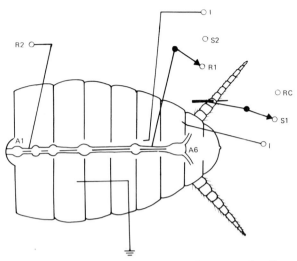

Figure 3-2. Diagram of electrode placement by Gammon (1977) in free-walking cockroach preparation. A1 and A6 are numbered abdominal ganglia, R1, R2, and RC are recording electrodes 1, 2, and the cercal recording electrode. S1 and S2 are stimulating electrode connections and I the indifferent electrode positions with stimulation. Recording was done with respect to the tissue ground (shown schematically). From Gammon (1977).

c. Electrode Placement

A watchmaker's eyepiece rather than a dissecting microscope was used for placing electrodes. This allowed measurably more freedom and the magnification of $2\times$ was sufficient to view the insect to determine the position of internal structures for electrode placement.

Silver wires were implanted through small holes in the ventral sclerites made by puncturing with a sharpened tungsten probe (cf. Part I, 9, a). During electrode placement, signals were recorded and electrodes were sealed in place when good signals were obtained in response to air puffs on the cerci or to shocks at the stimulating electrodes.

The electrode (Figure 3-2, R2) on the anterior aspect of the abdomen could be located in a straightforward manner since little or no tissue interposes between the cuticle and the ventral nerve cord. The posterior electrode (Figure 3-2, R1) was more difficult to locate since in this position the internal anatomy is slightly less straightforward.

Nervous impulses were recorded from wire electrodes insulated except for the flat tip and simply placed on the nervous tissue in question. From the size of the potentials obtained (Figure 3-3), it is evident that this method of electrode placement gives signals of good quality, comparable to other methods of extracellular recording.

The wax used to seal wires into the cuticle was a mixture of 1:1

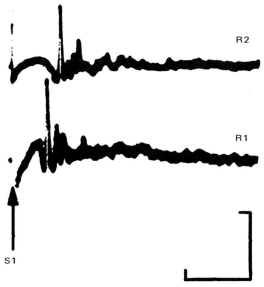

Figure 3-3. Traces recorded from electrodes R1 and R2 upon stimulation of the cercal nerve (S1) in the free-walking cockroach. From Gammon (1977). Calibration: 1 mV vertical; 5 msec horizontal.

beeswax and resin (Colophony, Paleamber, BDH Chem. Ltd). The wax could be worked by heating to the melting point and could be stuck to the cuticle as long as the area near the wire electrode entry was dabbed dry with tissue paper. Another small drop of wax was placed on each lead away from the implantation site to ensure stability. Then the various leads were gathered and wrapped together into a cable.

The cable passed out through appropriate slits in the experimental chamber, made of brass, and each lead was attached to a contact point. The internal dimension of the chamber was 20 × 40 × 60 mm, and the chamber rested on a temperature plate which employed the Peltier principle to maintain a constant temperature. Gammon's publication (Gammon, 1977) lists the experimental chamber as made of Perspex, 25 × 50× 60 mm.

d. Physical Positioning of Leads

Electrode connections for switching between recording and stimulating are somewhat instructive. Leads were kept as short as possible and shielded thoroughly. Leads from R1 and RC were taken to a simple metal connector box (Figure 3-4) where a connection was made to the center of a single-pole, double-throw toggle switch. The other two contacts of each switch were connected to either an Isleworth A-101 amplifier, or a home-

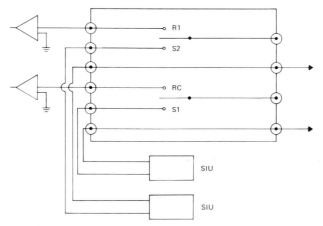

Figure 3-4. The physical arrangement of the switch-box used to switch between recording and stimulating electrodes as shown in Figure 3-2. SIU are stimulus isolation units, the arrows on the right are indifferent electrode connections, and the lettering is the same as Figure 3-2.

made stimulus isolation unit (SIU), and then a pulse source for stimulation. The SIUs were built from a Texas Instruments TIL lll light-emitting diode according to Figure 1-22.

3. Delcomyn Cockroach Preparations

a. Free-Walking or Tethered Preparation

For studies on the cockroach, either adult males or females were lightly anesthetized with CO_2 and the wings were removed. Insects were tethered to a stick by wax (Caulk sticky wax, L. D. Caulk Co.) and were given a styrofoam ball to hold upon waking (Delcomyn, 1973). The wax covered the dorsum and therefore the area from the pronotum to the terminal abdominal segments (Figure 3-7) was maintained rigid. Alternatively, the cockroaches were tethered to a line and allowed to walk freely, dragging recording electrodes (Delcomyn and Usherwood, 1973) (Figures 3-5 and 3-6).

For the freely walking or tethered cockroach, small holes were punched in the cuticle over or adjacent to specific muscles in the coxae, in this case the main extensors and flexors of the femur. The notation and anatomy of Carbonell (1947) are usually the standard sources for *P. americana* muscles. Small insulated copper wires (Belden 49 gauge, 35.6 μm diameter or 48 gauge, 40 μm diameter) were twisted together to form a cable, guided to a common point on the tether, and then inserted into the coxae

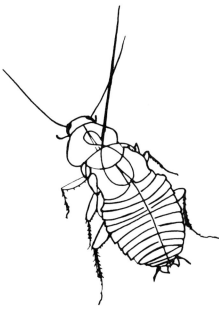

Figure 3-5. Details of tethering for recording from a free-walking cockroach, *Periplaneta americana*. Taken from Delcomyn (1973).

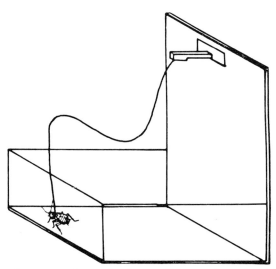

Figure 3-6. Arena for free-walking cockroach. After Delcomyn and Usherwood (1973).

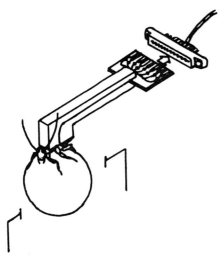

Figure 3-7. Physical arrangement of a tethered cockroach held over a styrofoam sphere for walking experiments. After Delcomyn (1973).

and anchored there with a small drop of wax. A thin piece of monofilament fishing line which was added to the cable prevented excessive twisting during the animal's movements (Figure 3-5).

The freely moving cockroach, so wired, was placed on a metal plate which served as a ground, and a barrier of greased Lucite sides prevented escape (Figure 3-6). For the tethered cockroach, so wired, cable and harness were not used.

Muscle potentials were amplified by Isleworth A-101 or Grass P9 or P15 amplifiers. The wiring diagram of instrumentation connections was shown in Delcomyn's article (Delcomyn, 1973) and is reproduced here (Figure 3-8). Two wires were inserted into the appropriate muscle and potentials were recorded differentially without a ground. A separate ground placed in the abdomen was useful in reducing interference from a position-detecting oscillator (Figure 3-8, O). Potentials were recorded for up to 3 days, but some clarity in recorded potentials was lost with time.

b. Tethered and Eviscerated Preparation

One preparation where recording from central neurons during free walking in the cockroach becomes a distinct possibility was recently described by Delcomyn (1976; Figure 3-9). Again either male or female adult *Periplaneta americana* were anesthetized with carbon dioxide. The insect was impaled on a pair of parallel No. 7 insect pins along the lateral margins of the abdomen and thorax as shown in Figure 3-9.

The dorsal part of the abdomen was removed from the segment anterior

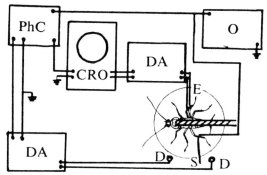

Figure 3-8. Block diagram of the instrumentation used in Figure 3-7 preparation. CRO, oscilloscope; DA, differential amplifiers; O, oscillator; PhC, phase converter of a position detector; D and S, detectors and sensors of the position detector; E, recording electrodes for myograms. From Delcomyn (1973).

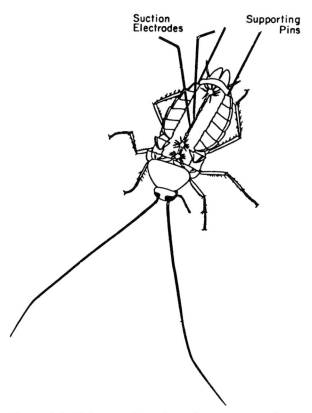

Figure 3-9. Eviscerated cockroach on a styrofoam sphere for free-walking preparation and central nervous recording. Suction electrodes are shown in place for central recording. From Delcomyn (1976).

to the entry point of the insect pins. The dorsum of the thorax was also removed up to the pronotal shield. After removal of the medial strip of the dorsal cuticle, visceral, fat body, and reproductive organs were removed. This exposed the ventral nerve cord which is accessible in the abdomen after removal of the ventral diaphragm.

Upon recovery from anesthesia, the cockroach was given a light styrofoam sphere to grip. Despite the severe surgery, the cockroaches behaved apparently normally in different activities including grooming and walking and responding to stimuli. Suction electrodes placed against the ventral nerve cord were used to record central nervous activity (cf. Part I, 2).

The method described is applicable to many other insects. Perhaps the remarkable feature here is that so much behavioral repertoire is left intact despite drastic dissection. Perhaps this merely confirms the localized and self-contained nature of the nervous system associated with each of the ventral ganglia.

4. The Tethered House Fly

Certain insect muscles are ideally suited for external recordings. The flight muscles of most Diptera are paramount in this category. Indirect flight muscles of the house fly *Musca domestica* are arranged vertically (the dorsoventral muscles) or more nearly horizontally (the dorsolongitudinals). Both types of indirect flight muscle are of the "fibrillar" category, i.e., unusually large multinucleate muscle cells.

a. Flight Muscle Location

The thorax of the house fly contains both a left and a right dorsolongitudinal muscle. Each muscle is composed of six fibrillar muscle fibers, sometimes referred to as "giant" because each cell is between 1 and 2 mm in length and 0.3 mm in diameter. The position of the dorsolongitudinals can be easily determined by considering the pattern of bristles on the dorsum of the thorax (Figure 3-10).

The scanning electron microscope photograph in Figure 3-11 shows the major dorsocentral bristles with prominent sockets. Three patterns of hairs are evident. Besides the major bristles there is an array of smaller hairs with prominent sockets, and a third series of hairs that lies close to the cuticle as a dense carpet. These smallest hairs are oriented to form four black stripes along the dorsum of the thorax. The stripes are black when viewed from the rear of the fly by the action of reflected light and the stripes change to a lighter color when light is reflected from the opposite direction.

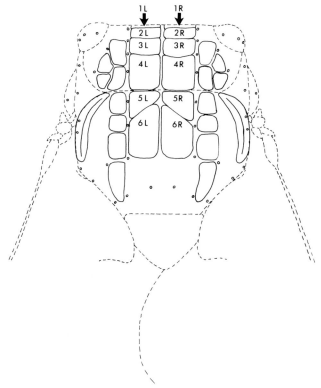

Figure 3-10. Dorsal view of thorax of house fly. The thorax and wings are shown as dashed lines. The anterior insertions of the dorsolongitudinal flight muscles are indicated by solid lines and labeled 1-6 L and 1-6 R, and the dorsal insertions of the dorsoventral flight muscles are solid lines, unlabeled. The small circles indicate positions of major bristles. The dorsal thorax is divided by two prominent horizontal sutures into the prescutum anteriorly, the scutum centrally, and the scutellum posteriorly.

The patterns of hairs and major bristles on the thorax of the house fly enable the underlying muscles to be located. The labeled outlines in Figure 3-10 show the anterior insertions of the dorsolongitudinal muscles. Thus if a pair of electrodes, even if they consist of rather large diameter wires, were inserted symmetrically between the left and right dorsocentral bristles and on the anterior part of the scutum immediately posterior to the transverse suture, the placement would be in the area of insertion of the top two parts of the dorsolongitudinal muscle. For purposes of identification the muscle fibers of the dorsolongitudinal muscles shown in Figure 3-10 are numbered 1 through 6 starting ventral most and are identical to the same muscles of *Calliphora* or *Drosophila* (Ikeda et al., 1976). Thus the dorsal-most pair of muscle cells are termed 6 right, 6 left, 5 right,

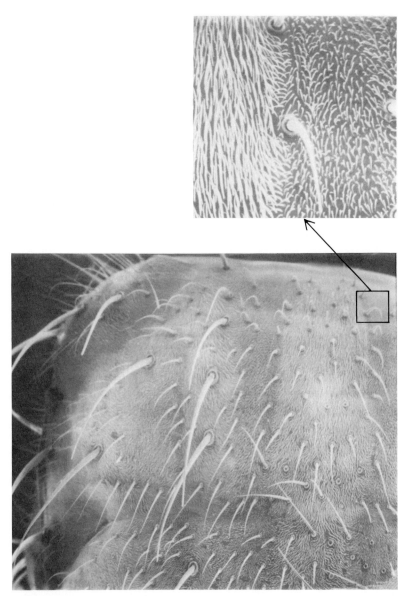

Figure 3-11. Scanning electron micrograph of the left anterior part of the dorsum of the thorax of the house fly. A shadow pattern is produced by the reflection of light from minute bristles next to the cuticle surface. The bristles are shown close up on an inset.

and 5 left or 6R, 6L, 5R, 5L. Unfortunately, Nachtigall and Wilson (1967) number the muscles I to VI starting dorsally, but others have not followed this system.

b. Mounting the House Fly

For mounting the house fly a miniature soldering iron (Oryx, Model 5S, W. Greenwood Electronic Ltd., 21 Germain Street, Chesham, Bucks., England) with a 4.5 V transformer was operated from a Variac® or other variable-voltage source (also used: Ohmitrol, Model PCA 1001, Ohmite Manufacturing Co., Skokie, Illinois 60076). A 30 AWG (254 μm) copper wire whisker was soldered to the S-type Oryx tip and allowed to protrude about 1 cm beyond the tip. By varying the voltage to the soldering iron, the temperature of the whisker was adjusted to just above that of the melting point of Tackiwax® (CENCO).

Upon touching a dish of Tackiwax, the wax formed a molten drop on the end of the whisker. This drop was then touched to the scutellum of an anesthetized house fly where it adhered readily and dried immediately. With care to avoid wetting adjacent structures, especially the alula or membranous lobes at the base of the wings (which lie near the scutellum at rest), a small mound of wax can be built upon the top of the scutellum (Figure 3-12). The head of a No. 5 insect pin (with the bead removed) was treated with wax and affixed to the scutellum of the fly by briefly interposing the soldering iron whisker between the lumps of wax until they melted together.

Once firmly attached to the insect pin, the fly may be allowed to recover from the anesthesia. Cold, CO_2, or ether anesthesia have been used, all with similar results. CO_2 is usually used for convenience. Hsiao (1972)

Figure 3-12. Photograph of a house fly with wires inserted and waxed into place for recording dorsolongitudinal flight muscle activity during walking or flying. The wax tether is just visible on the scutellum of the thorax.

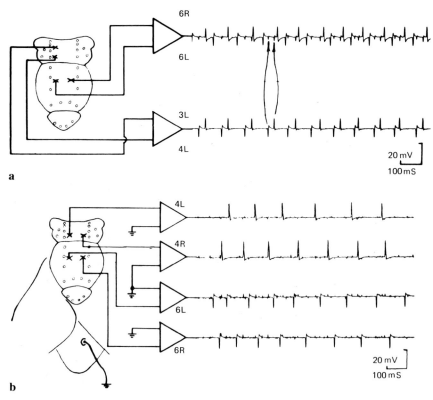

Figure 3-13. Wiring diagrams of the fly shown in Figure 3-12 connected for differential recording (a) or single ended recording (b).

used a slightly different technique to mount noctuid moths, *Trichoplusia ni*. He vacuumed the thoracic scales away, then waxed a nichrome wire to the thorax by passing a brief current through the wire, enough to melt a small drop of wax between the wire and thorax. A 2:1 mixture of beeswax and resin was used.

c. Electrode Arrangement

Electrodes were inserted by puncturing the cuticle with a sharpened tungsten probe (Part I, 9, a), inserting the electrode wire just into the wound, and sealing with wax (Figure 3-12).

The electrode arrangement may include two pairs of electrodes recorded differentially by two amplifiers (Figure 3-13a) or four separate signals may be recorded each with respect to the same common ground placed in the abdomen (Figure 3-13b). Figures 3-13a and b show wiring diagrams and potentials recorded from the flight muscles of the house fly.

Voltages recorded from the flight muscles are on the order of 20 to 50 mV or equivalent to intracellular potentials. Because the wires used for electrodes are of low impedance, the signals may be monitored directly on oscilloscopes or recorders without amplification, thus simplifying the recording procedure.

Another advantage is that the flight muscles are activated in an extremely regular pattern in normal flies and are activated at or below room temperature (25°C) when the fly is at rest. However, to record the muscle potentials during flight, the flies may be mounted on a phonograph cartridge (the cheap ceramic type for the largest signal is preferred) which, in turn, is mounted on the moving portion of a simple relay (Figure 3-14). A relay activated from a 9 V DC dry cell battery is most convenient and the voltage output of the phonograph cartridge may be recorded on a channel adjacent to the muscle potential recording (Figure 3-15). R. K. Josephson (personal communication, 1976) has found the Pixie transducer (Endevco) ideal for recording vibrations from singing in the katydid in the region of 200 Hz. For monitoring vibrations caused by flight in tethered moths, a strain gauge was used (Perumpral et al., 1972).

The procedures outlined above for the dorsolongitudinal flight muscles of Diptera are applicable to most other insects. Ikeda et al. (1976) have found that the dorsolongitudinal muscles (DLM) of *Sarcophaga bullata* and *Drosophila melanogaster* consist of six muscle cells inner-

Figure 3-14. Survey photograph of the house fly flight motor preparation. The fly is tethered to the needle of a phonograph cartridge and the cartridge is held by a 6-V DC relay which when activated pulls the fly from the wheel and induces flight reflexively. Recording wire electrodes (not visible) are connected to amplifiers by pinching the wires in microgator clips (shown at the top on the terminal strip).

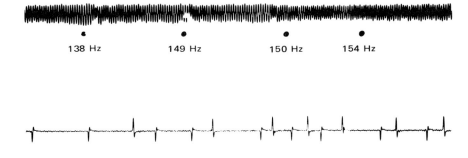

Figure 3-15. Asynchronous flight.   Simultaneous recording of phonograph cartridge (a) and flight muscle potentials (b) during flight as from Figure 3-14.   Note the much higher frequency of wing-beat vibration compared to the frequency of activation of the flight muscles.   Calibration mark: 160 msec.   Wing-beat frequency is measured at the dots indicated on the record to show slight variations in rate over 1 sec.

vated by five motor neurons. The two most dorsal muscle fibers in either the left DLM or the right DLM are innervated by axons which branch from a single motor axon. The same is true for *Musca domestica*.

Furthermore, Mulloney (1970) pointed out that in addition to Diptera the DLM of locusts (Neville, 1963), moths (Kammer, 1967), bumblebees and probably milkweed bugs *(Oncopeltus fasciatus)* each have five motor neurons innervating five separate bundles of fibers in each DLM. This implies a remarkable consistency in the Insecta and yields considerable order to comparative studies.

d. Alternate House Fly Procedures

The detailed description given above is only one method used for mounting and recording extracellular potentials (electromyograms) from insects, e.g., Diptera in this case.   There are many variations on this basic procedure.   Richard Hart (Wellcome Research Laboratories, Berkhamsted, England), for example, mounts two tungsten electrodes (Transidyne Model 404 or equivalent) together in a holder. He penetrates the dorsal thorax of *Lucilia sericata* flies in the area coinciding with 6R and 6L, described above, for *Musca*. He then glues the electrodes to the fly and suspends the fly using the electrodes as a holder.

When the fly recovers from the anesthesia, it normally begins and continues flight being out of touch with the substrate. *Lucilia* provides especially good flight muscle potentials compared to *Musca domestica*.

Forty-eight SWG (Ω27-μm) insulated copper, 25 μm stainless steel, and 25 μm platinum or silver or sharpened tungsten have been used for recording flight muscle potentials from Diptera. Each of these materials has its own peculiar advantages; copper and especially silver are flexible, and tungsten is stiff, for example. In short-term experiments of up to 8 hr, all of these metals give good potentials. For chronic recordings lasting several days, some workers prefer the precious metals.

In our hands flight muscle potentials recorded from house flies using bare copper were adequate for 1 day but tend to show deterioration on successive days; however, Derek Gammon and Paul Burt find copper wires give good nerve recordings from American cockroach for several days (1978, personal communication). The katydid implanted with 50 μm diameter insulated silver wires did not usually sing on the evening following the operation (Josephson and Halverson, 1971) but did sing on subsequent evenings. Thus an implantation would have to last at least a day in this case.

e. Antidromic Impulses

Mulloney (1970) adopted a method for producing antidromic nerve impulses to studies with Diptera, *Calliphora vicina (erythrocephala),* and *Eristalis tenax.* Fifty micrometer diamel-coated silver wires were inserted into a specific muscle in the dorsolongitudinal array. When shocked with a single 0.1 msec pulse, a nerve impulse was produced which was conducted from the muscle into the thoracic ganglion. The voltage to the stimulating electrodes was increased gradually until a response was observed from separate electrodes monitoring the flight muscle potentials.

This powerful tool enabled Mulloney to draw certain conclusions concerning the electrical connections in the central nervous system between single motor units in the right dorsolongitudinal muscle of *Calliphora* and *Eristalis.* It was reported that using a ground electrode with the stimulating arrangement described above reduced the stimulus artifact recorded by monitoring electrodes.

An important precaution to observe when interpreting results of antidromic stimulation was described by Mulloney and Selverston (1972). In some preparations antidromic impulses do not invade all branches of a given neuron to produce postsynaptic potentials at synapses known to exist with neighboring cells. Therefore failure to show interactions by antidromic potentials does not mean they are missing, only that they might be. A positive response from antidromic impulses is considered valid evidence for neuron interactions (Mulloney, 1976).

5. Cricket Preparations

Early work on stimulation of the insect nervous system, especially the brain, was conducted by Franz Huber (1960) who has continued research on neural integration over the intervening years (Huber, 1974). Huber and his colleagues have done much research with crickets and this includes a number of notable students and postdoctorals who have become infected with an interest in the neurophysiology of crickets and other related orthopterans.

The cricket song is a rather stereotyped sound which is species specific and comes in three varieties, the calling, courtship, and rivalry versions (Kutsch, 1969). Because the male cricket is normally stationary near an identified territory when singing, he represents a convenient choice for studying the underlying neural control for the descrete behavior represented by the singing.

a. Brain Stimulation

After Huber's preliminary experiments, Rowell (1963a) described the parameters and problems encountered when trying to deliver shocks to the nervous tissue of locusts. Moreover, Rowell (1963b) described a method of implanting stimulating electrodes into the brain of the locust, *Schistocerca gregaria,* which was freely moving. The first of these studies showed that the current of flow actually delivered to the tissues depends on the frequency of stimulation and is not readily predictable. At higher voltages or higher frequencies of stimulation, gas generated during polarization of the electrode tip can produce distortion of the tissues with subsequent further alteration in the properties of the electrical coupling between the tissue and the electrodes.

Dietmar Otto (Zoologisches Institut der Technischen Hochschule, Darmstadt, Germany) has continued development of methods for shocking the brain of the cricket using free-walking preparations (Otto, 1967, 1971; Kutsch and Otto, 1972). Insulated steel wires are obtained from Electrisola of 20 μm diameter. Two basic electrode arrangements are possible, two electrodes mounted together in the brain to be shocked, or a single electrode mounted in the brain with an indifferent electrode placed in the hemolymph at some distance away, usually the thorax.

Mounting two wires (bipolar electrodes) provides the greatest localization of shocking current; however, this is achieved at the expense of a larger entry wound and single sharpened electrodes are easier to place. In addition, bipolar electrodes normally do not produce good responses unless placed some distance apart.

Stimulation usually employs 5 to 40 μA of current at variable pulse widths. As a general rule 0.1 msec produced no bubbles, but shocking pulses longer than 0.5 msec did produce electrolytic gas at the electrode tip (Otto, 1973, personal communication).

b. Suction Electrode Recording

Recording from the freely moving cricket has been accomplished by a rather elaborate suction electrode procedure (Stout, 1971). A polyvinyl chloride tubing of 0.6 to 0.9 mm inside diameter was drawn to a fine tip over a small flame. The tip was cut to leave an inside diameter of down to 20 μm. The tubing was trimmed to 2 cm and pushed onto a short piece of a stainless-steel needle taken from a 20 gauge syringe needle. A 40 μm stainless-steel wire was inserted 1.5 cm into the barrel of the plastic tubing then bent over the open end of the needle. A 60 cm piece of plastic tubing was pushed over the free end of the needle tubing with the steel wire protruding from the joint, and Uhu®cement [UHU-Werk, H.u.M. Fischer GmbH, 7580 Buhl (Baden) Switzerland], a general purpose glue similar to DUCO cement, was applied to the joint for an airtight seal. Suction was provided by a 5 or 10 ml syringe.

Stout (1971) gave a detailed description of electrode placement. The procedures suggested had to be preceded by attachment of a copper collar to the cricket (Figure 3-16). All attachment and cementing was done with a beeswax–colophonium mixture. This could be melted and molded so that the collar was rigidly affixed to the pronotum immediately in front of the prothoracic legs.

After construction of the electrode and collar attachment, crickets were immobilized and turned to expose the ventral area of the neck connective. The cuticular membrane was torn exposing the ventral nerve connectives.

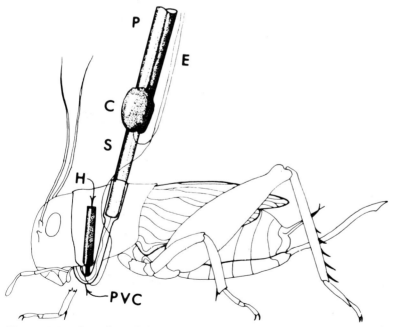

Figure 3-16. Drawing of the suction electrode mounted to record activity in the ventral nerve cord of the cricket. H, harness; PVC, polyvinylchloride tubing; S, syringe needle tubing; C, cement; P, flexible lead tubing; E, electrodes. After Stout (1971).

A temporary glass rod-holder was waxed to the suction electrode, and the electrode was manipulated to lie against one of the ventral connectives while also lying adjacent to the horizontal bar of the copper collar.

With the electrode against the connective and a vacuum applied, nerve impulses of 200 μV to 2 mV were recorded differentially with respect to a second stainless-steel wire placed in the hemolymph near the connective. When an appropriate pattern of nerve impulses was obtained, usually nervous impulses from some of the larger diameter axons in the connectives, the suction electrode was waxed to the collar, the temporary holding rod was removed, and the suction and indifferent electrodes were waxed to the sides of the collar to end in a position pointing vertically away from the upright cricket.

The neck wound was sealed with the suction electrode and indifferent wire waxed into the hemocoel and the cricket was freed. Both electrode leads along with the plastic tube used for suction were led to the recording amplifiers and the vacuum syringe in a position over the cricket.

The mounted electrodes added a burden of 80 mg to the cricket. However, this had no apparent effect on normal behavior. With the electrodes in place, the crickets walked freely, courted, and copulated.

One of the most conspicuous units in each of the locust ventral connectives is the descending contralateral movement detector axons (O'Shea et al., 1974). Since the records of Stout (1971) show responses to movements in the visual field, the conspicuous cricket axonal activity is no doubt equivalent to the locust unit. In fact this same unit may also be the conspicuous axon profiles seen in diptera neck connectives (Coggshall et al., 1973) where they evidently function in the start of flight (Mulloney, 1969).

Since these prominent axons, the descending movement detectors, occur from Orthoptera to Diptera and are involved in response to visual stimuli, they may be a common feature in other insects as well. It would be interesting to examine their possible interaction with the ascending "giant" interneurons which are so conspicuous in the posterior ventral nerve cord of many orthopteroid insects, but whose exact role is less clear (Parnas and Dagan, 1971).

6. Notes on Electrode Placement

There are a few hints which assist in electrode placement during work-up in freely walking preparations. Burns (1973), for example, points out that since the tarsi of locusts are devoid of intrinsic muscles, wire electrodes implanted there are most certainly registering sensory nervous impulses from a variety of sensory structures known to be present including trichoid hairs, chordotonal organs, and campaniform sensillae.

a. Audio Monitors

Burns listens to his recorded signals by taking the amplified potentials through an audio amplifier. It has long been known that the human ear is extremely sensitive to the "pops" and "buzzes" which represent the audio equivalent of nervous impulses. Far more can be detected by listening to nervous impulses than by watching recorded potentials on an oscilloscope trace.

Furthermore, the audio monitor frees the experimenter's hands and eyes to manipulate recording wires for placement to produce the best recording. Gammon (1977) probed his cockroaches with the recording wires until the best signal was obtained. Perhaps the most remarkable thing about Gammon's accomplishment is that he used only the cut ends of straight wires as the recording surface, whereas Runion (Runion and Usherwood, 1966) scored an insulated wire then wrapped the scratched insulation against the appropriate nerve bundle.

It may be argued that Runion was recording a smaller voltage than Gammon, for the cockroach ventral nerve cord produces comparatively large signals. In other cases where the activity must be recorded from small peripheral nerves, perhaps more care might be needed to obtain the necessary signals.

Just as Delcomyn found that the cockroach could withstand extremes in dissection, the number of electrodes implanted does not appear to be limiting. Elsner (1968 a, b) has recorded the neuromuscular events during grasshopper courtship of *Gomphocerippus rufus* (L.). Using 30μm steel wire electrodes, Elsner reported that *G. rufus* males conducted apparently normal courtship behavior with up to 37 electrodes implanted in various muscles of the thorax (Elsner, 1976, personal communication). Thus numerous units may be monitored if necessary to map the motor activity used during complex behavior.

b. Adhesives

One major problem in free-walking preparations is the adhesive or cement which is used to seal the electrode through the cuticle and to anchor the trailing cord of electrode leads. Several materials have been employed for this purpose. Some are better for one insect and useless for others.

Delcomyn used a Caulk sticky wax (L.D. Caulk Co.) and Gammon used a mixture of beeswax and resin (Colophony, Paleamber, BDH Chem. Ltd.) for adhering to the cuticle of the American cockroach. The mixture of beeswax and resin has been particularly popular in Germany for cementing electrodes in crickets (Part I, 9,f). Pickard and Welberry (1976) used cyanoacrylic adhesive to seal honey bee cuticle (Eastman 910®, Zip-Bond®, Bond-Solv®, Bond-Fix®, Reid Products, Tescom Corp.).

Various dental waxes have been used to seal electrodes into cuticle. Ewing and Manning (1966) used dental wax for recording from the cockroaches *Gromphadorhina portentosa, Nauphoeta cinerea,* and *Periplaneta americana* but did not specify a source or type.   Palka (1969) used acrylic dental wax to hold the head of the cricket *Acheta domesticus* to the thoracic shield.   Runion (Runion and Usherwood, 1966) found the fast-drying Eastman 910 adhesive very useful on *Schistocerca gregaria* cuticle for sealing cuticle or adhering wire electrode leads.   Corning et al. (1965) used prosthetic cement (Nu-Weld) to cover screws in *Limulus polyphemus* (horseshoe crab) cuticle and provide a support for implanted electrodes.

Tackiwax® (CENCO) readily adheres to any cuticles such as adult or pupae Diptera, *Musca* sp., *Phormia* sp., and *Calliphora* sp., but is ineffective alone for adhering to cockroach cuticle, *Periplaneta americana,* or to maggot larvae. Dentina ribbon wax (Amalgamated Dental Co.) is similar to Tackiwax. A mixture of 52°C paraffin wax and petrolatum at 3:1 has been used to seal tubing into the spiracles of cecropia pupae (Schneiderman and Schechter, 1966).

The selection of waxes for adhering to insect cuticle is an inexact science.   To determine which wax or adhesive is best for a particular application almost requires a haphazard search of available materials.   This usually leads to the departmental cupboards and almost always is poorly reported in research publications, if not ignored.

Insect cuticle, of course, is not a static material. The waterproofing external layers of the cuticle of terrestrial insects may be actively maintained, and it is thought that the physical properties of the cuticle may change under hormonal influence. Materials may be exuded onto the surface of the cuticle from epidermal glands thereby possibly reducing the adhesion of externally applied cement or wax.

c. Electrode Localization

Stainless-steel electrodes have been preferred for stimulation because of their amenity to localization of the recording site by deposition of iron and conversion to Prussian Blue.   The following procedure is a general description of the method of marking the stimulation site.

The steel wire used for recording is made the anode for a current of about 2 μA which flows between the recording electrode and a tissue ground electrode. After the current flows for 15 to 30 sec, a small amount of iron is lost from the stainless-steel surface. The tissue is then fixed in 10% formalin containing 1% potassium ferrocyanide.   The reaction produces a Prussian Blue spot at the site of the electrode tip (Green, 1958; Grundfest et al., 1950).

Rowell (1963b) also used the iron deposition method but fixed the tissues with the wire electrodes in place first. A 120 V DC battery then was connected in series with a 2.8 MΩ resistor, a microammeter, and a pair of metal forceps. About 40 μA of current was passed from the forceps through the implanted steel wire electrode and into *Schistocerca* brain with the electrode made anode. The return current path was provided by a carbon rod, and the current was passed for 10 to 20 sec. A second hour period of fixation allowed the deposited iron to diffuse away from the electrode before soaking the tissues in potassium ferrocyanide for an hour.

Other methods for marking the recording site of metal electrodes were described by Bentley and Kutsch (1966). When 100 μm nichrome wire was used for recording, crickets were anesthetized with CO_2, and the thorax was removed with the electrode wires in place. The thorax was perfused and soaked for 12 hr in a solution containing 5% sodium acetate, 1% dimethylglyoxime, 5% ammonia, and 87% of a solution of 47% ethanol. After this treatment, current was passed through the wire with the recording electrode made the anode. A bright red nickel–dimethyl-glyoxime complex was precipitated around the electrode tip thus marking the muscle of interest.

Josephson and Halverson (1971) resorted to an even simpler procedure for marking the tissue around the tip of a recording electrode. Because of a lack of readily available steel wires, silver was used instead. After recording the behavioral events of interest, the katydid, *Neo-conocephalus robustus*, was anesthetized with CO_2. A current (100 μA for 10 sec) was passed between each of the recording electrodes (made the anode) and the reference electrode. The katydid was then fixed in a solution made of 10% formaldehyde and Kodak D 19 photographic developer in 1:1 combination.

The silver deposited by the current flow was reduced by the developer leaving a small black spot to mark the position of the recording electrode. This procedure is simple and has the advantage of using silver wire which is slightly better than copper for long-term recording and more readily available than steel. The developing was done in a lighted room.

d. Hall Generators

Hall generator devices have been used in physics for some years to study the Hall effect. When a current flows between two opposite edges of a thin rectangular piece of metal, a voltage drop can be measured across the remaining edges when a magnet is brought near the surface of the strip of metal. This is the Hall effect and the voltage induced depends on the distance between the magnet and the surface of the Hall device.

Some Hall devices are sufficiently compact to be mounted on larger insects. When a magnet is also mounted on an adjacent limb or structure which moves with respect to the Hall device, precise measurements of movement using the Hall effect can be obtained. This method has been applied to the measurement of ventilation movements in locusts, and limb and wing movements in crickets during singing or walking (Hustert, 1974, 1975; Elsner, 1970, 1974a,b; Möhl, 1972; Naynert, 1968).

Some Hall generators are sufficiently compact to be used on larger insects for measuring abdominal movements during ventilation in locusts, and the movement of limbs and wings during singing in crickets. Dr. Reinhold Hustert (personal communication while Dr. Hustert was in Seattle, 1975, now at Fachbereich Biologie Universität Konstanz, 775 Konstanz, Germany) kindly provided the following particulars from his experiences with Hall generators: Type SBV 566 (U.S. $5.00, 1975) Hall generator weighing 0.2 g was obtained from Siemens Corporation in New Jersey. This model contains ferric elements which slightly attract the magnet used. Another model available from Siemens, No. SV210, costs several times more (U.S. $35.00 1975) than the former, but is nonferric.

The Hall generator is supplied by 6 V DC from a dry cell battery and uses about 75 mA. The output voltages depend on the magnet used but are typically millivolts. The output can be measured of a Tektronix 502A oscilloscope or other DC amplifier sensitive to the low-millivolt range.

The Hall generator itself may be glued or waxed to insect cuticle. Hustert used a 2:1 mixture of beeswax and resin (colophonium or violin wax). Hustert cautioned that the wax be used under conditions in which stress to the insect is absent. Evidently expulsion of fluids can occur via the cuticular pore canals when some insects are exposed to stress.

A Krupp Koerflex 300 magnet was used of 300 μm diameter, however this model is no longer produced. The magnet is mounted to move either parallel or perpendicular to the plane of the Hall generator. Excursions of 20 μm can be measured without difficulty and the measurement is complicated by a nonlinear response characteristic of the device (Figure 3-17).

As shown in Figure 3-17, a magnet nearer the Hall generator produces a greater voltage change during excursion across the surface. Other variables not shown include the strength of the magnet which would also alter the response proportionally. For practical use, the Hall generator and accompanying magnet are arranged to operate in the linear range of the movement–response curve on either side of the midline.

Insulated copper wires of 50 to 100 μm are used for connections to the Hall generator. Four are necessary, two for supply voltage and two for voltage response.

When using the Hall generator, any nearby magnetic materials may interfere, and these must be isolated. It is convenient to glue the back of the Hall generator to a surface of interest; however, Dr. Elsner (Zoologisches Institut der Universität zu Köln, Germany) attached his devices by gluing the lead wires to the cricket's cuticle. The leads had been wrapped and were therefore rather stiff.

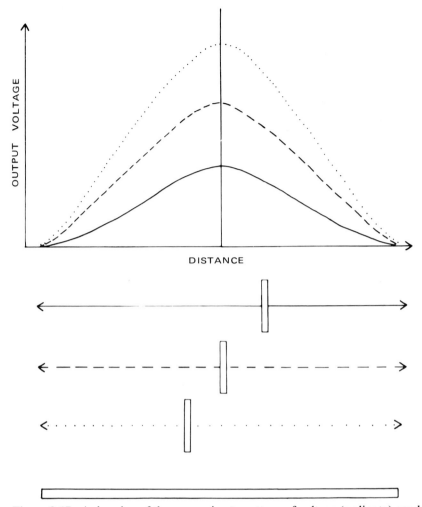

Figure 3-17. A drawing of the approximate pattern of voltage (ordinate) produced by the Hall generator when a magnet is moved (abscissa) parallel to the generator device. The dotted line is a magnet movement parallel to the Hall device at a short distance from it, the dashed line is at an intermediate position, and the solid line is farthest away from the surface of the Hall generator. After Hustert (1976, personal communication).

An example of an attachment of a Hall generator for recording locust ventilation movements is shown as reproduced from Hustert (1975) in Figure 3-18. The Hall generator devices have largely been replaced by newer methods in studies on leg movement (Elsner and Popov, 1978).

Several other devices have been described for measuring movements. Optical devices were used to measure ventilatory movements, spiracular opening movements, and heartbeat all from intact insects. These devices are described in Part IV under transducers.

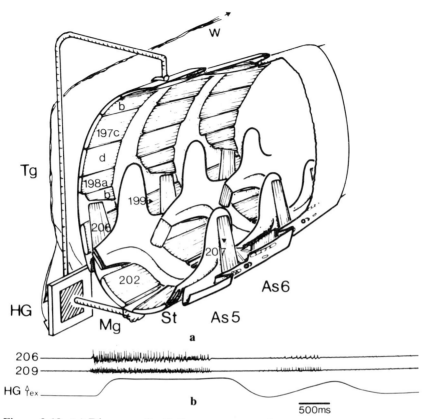

Figure 3-18. (a) Diagram of a Hall generator attached to the abdomen of a locust to detect respiratory movements. HG, Hall generator; 206, 207, and 209, muscles involved in respiratory movements; Mg, magnet. After Hustert (1974). (b) Recording from muscles 206 (top trace) and 209 (middle trace) and from the Hall generator (HG) during unrestricted respiratory movements of locust. After Hustert (1975).

e. Printed Circuit Electrodes

The use of the printed circuit technique in the construction of patterns of exposed metal surfaces on an insulated chip was reported recently by Pickard and Welberry (1976). The chip was placed either on the brain or beneath the central nervous sheath of the brain of a honey bee before sealing the head capsule and leaving a protruding cable of wires connecting external amplifiers with the implanted chip.

The chip was used to record brain activity as a surface recording electrode from freely moving and flying honey bees. The records obtained reportedly had characteristics similar to those obtained with glass, stainless-steel, or tungsten microelectrodes used to probe directly into the brain for extracellular recording (Vowles, 1964; Kaiser and Bishop,

1970). The printed circuit technique offers considerable promise in investigating central nervous activity during normal behavior in insects.

f. Wing-Beat Detector

Dr. Uwe Koch (M-P-I für Verhaltensphysiologie, D-8131 Seewiesen, Germany) has recently developed a sophisticated method of recording the position of insect wings during tethered flight. Previously only high-speed cinematography was able to render the tilt of a wing during flight.

The new method employs induction effects from a fixed external magnetic field detected by a sensing coil. The amplitude of voltage induced in sensing coils is roughly linearly related to the angle of the sensing coil with a magnetic field when the coil is oriented between 60° and 120° to the field. By using two sensing coils mounted on the same surface and oriented at 90° with respect to one another and the magnetic field, the exact tilt of the surface in a uniform magnetic field may be detected.

Koch (1978) used 18 μm diameter insulated copper wires (Electrisola) to fashion coils of 40 turns which weighed 0.6 mg. These were mounted on the wings of *Locusta*. Figure 3-19 shows sense coils mounted on the wings of a locust. U-D is nearer the body to detect up and down movements. P-S is located distally and oriented to record pronation and supination of the wing.

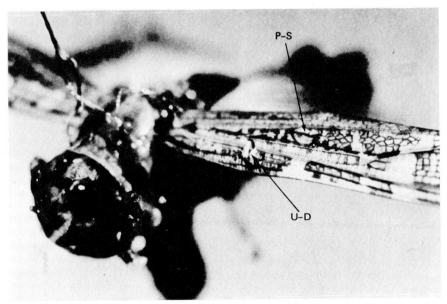

Figure 3-19. Three-quarter frontal view of a locust with wire coils mounted on the left wing. A portion of the lead wires is seen directly above the head. After Koch (1978, in press).

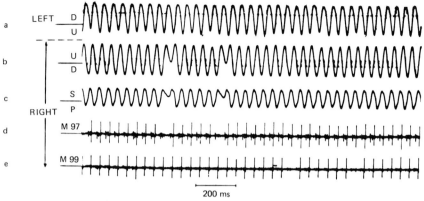

Figure 3-20. Recording of emf induced in the coils similar to those shown in Figure 3-19 during locust flight. (a) Down–up movements (D–U) of the left wing. (b) Up–down, U–D, (c) supination–pronation movements, S–P, of the right wing, respectively. (d) Recording from flight muscles M97 and (e) muscle potentials from M99 of the right wing. Note that when muscle M97 failed to fire, a marked change in wing position of only the right wing was produced; whereas, failure of muscle M99 caused no obvious change in the position of the right wing during flight. After Koch (1978).

Potentials were recorded from the right direct flight muscles M97 and M99 (Figure 3-20d and e) and potentials were recorded from sensing coils mounted on the right wing (Figure 3-20b and c) and on the left wing (Figure 3-20a) of the locust. During tethered, straight flight, nerve impulses failed during two separate cycles of the normal burst driving muscle M97 (Figure 3-20d) and the position of the right wing was altered for one wing-beat cycle; however, the opposite wing was not affected (Figure 3-20a). When a similar failure occurred during one cycle of the firing pattern of muscle M99 (Figure 3-20e), the wing position was much less affected.

The induction method of recording movement changes may have application to other movements. Its usefulness to the evaluation of insect flight is great since now large numbers of consecutive electrical and mechanical events may be analyzed rapidly in multiple-channel recordings.

Part IV

Organ and Tissue Preparations

1. Introduction

As a teacher of insect physiology, I have long admired the rich sources of materials including laboratory manuals, atlases, and descriptions of experiments which are available to students of vertebrate and invertebrate physiology with the exception of insects. Much of the biomedical equipment available is adaptable for use with frogs, rats, and crabs and is not easily adaptable to most insects.

Insect morphology references include the 1935 text by Snodgrass plus his various miscellaneous publications in the Smithsonian collections. Duporte's *Manual of Insect Morphology* (published by Reinhold, New York, last reprinted in 1961 by Litton Educational Publications, Inc.) is now out of print. Fox and Fox (*Introduction to Comparative Entomology*, Reinhold, 1964) is thorough but aimed at a more general level and *A Laboratory Manual for Insect Morphology* (Strickland, Hocking, and Ball, 1958, New York Scholar's Library) is just a brief introduction and poorly illustrated.

The existing material which could be used for anatomy or morphology reference in neurophysiology studies is scattered throughout the literature and is not pulled together in any one place with a few notable exceptions. There are rumors that Duporte has written an insect morphology text that may be published in the future. Perhaps publication of texts in the area of insect anatomy and morphology has not been actively sought by publishers.

The greatest detail of description on almost any insect from a neurophysiological or visceral organ standpoint pertains to grasshoppers, locusts, and cockroaches in terms of numbers of muscle fibers, innervation, and location (Albrecht, 1953; Guthrie and Tindall, 1968; Uvarov, 1966). All other insects are represented in volumes based mostly on biology and life habits. The latter include descriptions of *Drosophila* by Demerec (1950), ants by Wheeler (1910), housefly or Diptera by West (1951), West and Peters (1972), Hewitt (1914), Lowne (1890-1892), and Clements (1963), honeybee by Snodgrass (1956) and Dade (1962), and Lepidoptera by Kuznetsov (1915) and Portier (1949).

Despite a fairly abundant literature, albeit scattered, on the gross anatomy of insects, very few descriptions may be found of experiments that can be done with wet tissues which are useful for teaching principles of insect physiology. Perhaps the field has been moving so fast over the past two decades that it hasn't seemed appropriate to sit down and work out a few laboratory routines.

The one great exception to this is the demonstration of fast and slow muscle contraction which can be shown so beautifully with grasshoppers. Both Clark (1966) and Hoyle (1968) have descriptions of this procedure.

In gathering information for this book, I have come across laboratories where excellent preparations have been developed to demonstrate one or another principle in insect physiology. These preparations are ordinarily evolved by certain people and not generally known. There is an especially high barrier to exchanges of information of this sort between continents, but illustrative materials of use in teaching laboratories in the field of entomology may be quite different in two institutions located in adjacent cities.

The American Physiology Society recently began publishing *The Physiology Teacher* to increase the exchange of physiology information useful in teaching. Volume 1, No. 1 appeared in April 1971 and Volume 5, No. 3 is dated July 1976. Further information is available from the American Physiology Society, 9650 Rockville Pike, Bethesda, Maryland 20014. Although there have been a few articles in *The Physiology Teacher* (TPT) on experiments using insects which are extremely valuable, the main thrust of TPT remains vertebrate physiology, with emphasis on mammals. Comparative physiology covering subjects in the arthropoda receives some attention but the American Physiology Society cannot be expected to cover all fields and TPT is a step in the right direction.

The Kerkut series *Experiments in Physiology and Biochemistry* was started as Volume 1 in 1968 and apparently ended with Volume 5 in 1972 (edited by G. A. Kerkut, published by Academic Press). It also contains numerous useful examples of insect preparations and procedures. Included are tips on rearing, trouble shooting, lists of equipment needed, and valuable sections on avoiding problems.

The following sections outline several useful preparations. While it would be far better for individual authors to describe their own techniques, this is not done here in the interests of brevity and because author-contributed experiments are difficult to organize logistically. The examples shown below are meant to introduce particular ways of doing things as a starting point for development of other preparations in basic research and a starting pointing for laboratory experiments.

2. Transducers

The history of transducers used in insect physiology is similar to that of the development of insect actographs, and both are heavily dependent upon technological advances. Transducers may be roughly categorized as based on mechanical, photo, temperature, electric, or electronic principles or combinations of two or more of these principles.

a. The RCA 5734

The RCA 5734 tube contains a movable anode. When properly connected, small forces are recorded as voltage changes. There are a variety of circuits published (Lang, 1972; Lion, 1964; Geddes and Baker, 1975) and unpublished. The bridge circuits described by Lang (1972) and by P. L. Miller (personal communication, March 1976, Zoology Department, Oxford University, Oxford OX1 3PS, England) are instructive (Figure 4-1).

Lang used a 6-V storage battery for filament current and three Burgess V60 batteries at 90 V DC each in series to provide the plate voltage of 270 V DC. The RCA 5734 was held in a femur clamp to provide a heat sink and an insect pin was used to attach the tissue, *Limulus polyphemus* heart, to the anode (Figure 4-1a).

Peter Miller's circuits (Figure 4-1b and c) provide simple variation for the circuitry and show connections to the obsolete Tektronix 160 series power supplies which were popular several years ago and probably still available in established physiology or zoology departments. Miller used the RCA 5734 in his early studies on ventilatory movements (Miller, 1971a, b, c).

b. Strain Gauge

The Grass Instrument Co. strain gauge transducer, FT .03C, is rated in the range of 2 mg to 2 kg. The 0.1 to 10 mg range is of interest in insect physiology as this range covers extremely delicate movements encountered in some visceral muscles. The FT .03C then is useful for skeletal

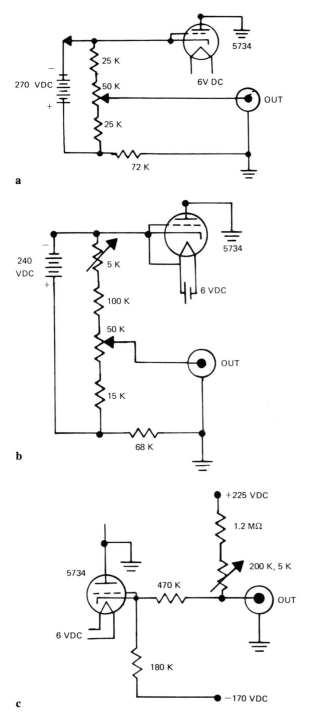

Figure 4-1. Circuit diagrams used with the RCA 5734 transducer tube. (a) Bridge circuit from Lang (1972). (b) Bridge circuit used by Peter L. Miller (unpublished). (c) A circuit powered by the Tektronix 160 series power supply which provided +225 V DC and −170 V DC.

muscles in insects with the precaution that self-resonance occurs at 85 Hz in the most sensitive range and the maximum displacement is 1.1 mm.

Some workers have developed their own strain gauges. McCann (1969) used a Statham transducer and an RCA CA 3010 operational amplifier to produce a sensitivity of 0.5 mg/mV (Figure 4-2). The circuitry includes adjustment for static load conditions. The actual transducer is composed of four 330 Ω strain gauges mounted in a housing (Figure 4-2). When connected as shown, the transducer is attached to the tissue of interest, then adjusted to a slack tension of zero. At this point (unloaded) the zero adjust potentiometer is set to bring the recorded potential to an arbitrary zero point on the recorder. Now the transducer is adjusted to develop rest tension or static tension and the recorded potential is readjusted to the zero level with the static load potentiometer and switch.

A similar strain gauge transducer may be constructed of foil strain gauges (BLH) by gluing four matched gauges to a strip of spring steel feeler gauge metal or some other metal sheeting. Perumpral et al. (1972) used a 0.015 x 0.5 x 4 in. brass sheet (0.381 x 12.7 x 101.6 mm) and showed details of mounting. A source of concern is that the frequency recorded is not near the resonant frequency of the device. This will depend somewhat on the loading characteristics and will be different for each mounting arrangement.

I have used four matched BLH type FAE-50-35S6L gauges of 350Ω and gauge factor 2 for physiological recording mounted on a 2 mil (0.002 in., 50.8 μm) thick feeler gauge (the tool which measures the gap in spark plugs). The latter device was slightly more sensitive than the Grass FT .03 transducer but was still too stiff for use with cockroach ventral diaphragm or heart movements.

BLH also makes semiconductor (silicone) strain gauges (Piezoline strain gauges). The advantage of semiconductor material is in the much greater gauge factors compared to wire or foil gauges. The gauge factor

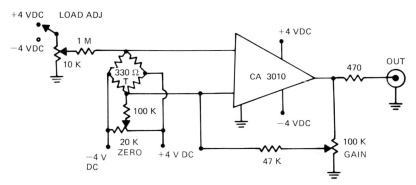

Figure 4-2. Amplifier circuit for four matched foil strain gauges connected in a bridge arrangement. After McCann (1969).

for foil is usually in the range of 1 to 4. Equation (1) shows the relationship between output

$$E_{out} = E_{in} \times \epsilon \times GF'$$ (1)

voltage, E_{out}, and the bridge input or exciting voltage, E_{in}, the strain ϵ (in microinches per inch), and the approximate linear gauge factor, GF', for a full bridge on a bending beam. Since the nominal gauge factor for piezoresistivity devices usually falls in the range between 100 and 150, an immediate improvement in sensitivity of the device is realized. Silicone strain gauges have excellent stability and high resistance, although they are somewhat fragile until mounted and are rather expensive.

Silicone strain gauges are also available from PYE Dynamics Ltd., Endevco, and Akers Electronic Co. The Pixie® transducer silicone strain gauges have been used in numerous applications (Sykes et al., 1970). Two examples are described.

Don Graham (1973, then at the Zoology Department, University of Glasgow, Scotland) used the Pixie Model 8101, range 900 to 1200 Ω, in a two-element bridge for a force transducer (Figure 4-3). The vital statistics for this physical arrangement shown are: 10 g cause 10 μm movement or not quite isometric; the linear range is 0 to 10 g at 50 mV/g; above 40 g the element can fracture; noise 10 μV; 55 μin./g compliance; draws 20 mA. Connecting two elements in opposing arms of the Wheatstone bridge reduces the temperature drift. The noise level makes this device of practical value down around the area of 1 or 2 mg.

R.K. Josephson (personal communication, 1976, Developmental Biology Department, University of California, Irvine, California 92717) de-

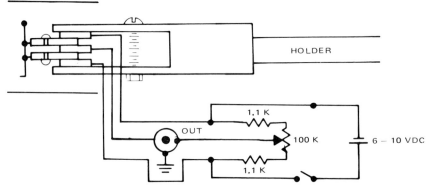

Figure 4-3. A diagram of the physical arrangement of a two-stage transducer using the Pixie® model 8101. The Pixie elements are mounted back to back as shown in two arms of a bridge circuit. The elements are squeezed by a machine screw in an inert housing (e.g., plastic), and the other elements of the bridge circuit are shown below powered by 6–10 V DC. After Don Graham (personal communication, 1973).

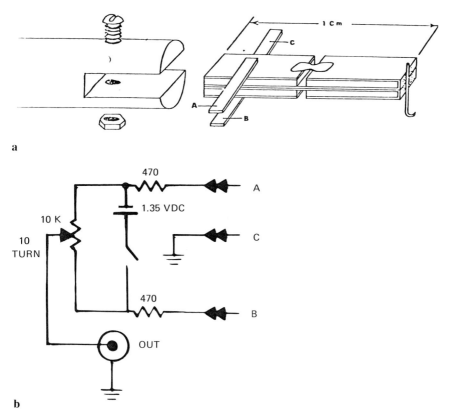

Figure 4-4. A second version of the silicon strain gauge transducer using Pixie® elements. (a) An exploded view of the gauges mounted back to back and a simple holder. A hook is waxed to the end for tissue attachment. (b) Circuit diagram for use with the transducer driven by 1.35 V DC. A, C, and B are terminal connections between the elements and the bridge circuit. After R. K. Josephson and P. L. Donaldson (unpublished, 1976).

scribed another similar version of the Pixie Transducer (Figure 4-4) at the XVth International Congress of Entomology (Electrophysiology workshop, August 22, 1976, Washington, D.C.). In this case, the transducer was designed to pick up the vibrations of Katydid singing and had to operate in the 200 Hz range.

Both of the examples described show a wealth of potential for the silicone strain gauge as adapted to insect physiology. Since both foil and silicone strain gauges use bridge circuits for maximum sensitivity, the bridge amplifier circuit of Dobkin (1973) may be of interest (Figure 4-5). The amplifier is nulled to zero output voltage by shorting pins 2 and 3 of LM 321, and then the bridge is connected and the voltage output again set to zero by adjusting the bridge zero potentiometer. The values of resistors in the bridge can be about the same as the strain gauge resistance and

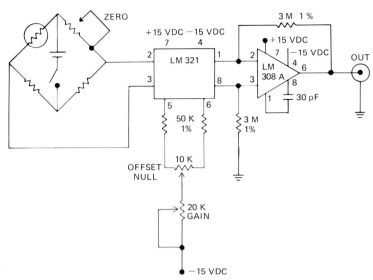

Figure 4-5. Generalized bridge circuit for use with resistance elements in measurement amplifiers. After Dobkin (1973).

more gauges may be added to the arms for increased sensitivity. The value of B1 will determine the bridge current and is selected to lie within the power rating of the bridge resistors and strain gauges.

If the circuit is needed only for low-frequency measurements, the frequency compensation capacitor (Figure 4-5, 30 pF) may be increased to 300 pF which also produces lower noise and greater stability. The gain adjustment may be accurately set at 1000 to calibrate the amplifier voltage output. The amplifier can measure voltage changes in the microvolt range, but sensitivity of measurement depends on the response of the strain gauge used, and therefore the gauge factor.

Recently Brandt et al., (1976) introduced a transducer capable of measuring forces in the range 1–10 mg. The device was adopted from a commercial transducer (AE 800 series, Akers Electronics, Horten, Norway) with a resonant frequency at 7 kHz. A circuit of bridge amplification was also provided plus a filter circuit. However, the unfiltered output is shown grounded on the figure in the publication (Brandt et al., 1976), when in fact the connection is not grounded.

Hsaio (1972) used BLH semiconductor strain gauges No. SPB-35-500 mounted in four matched pairs on two thin strips of phosphor bronze. The strain gauges were connected in a standard bridge circuit but two amplifiers were used to measure force and torque separately. The transducers could record 50 mg changes in force and had a rather high working noise level of about 10 mg from flight vibrations according to the published figures.

Foil or silicone strain gauges are useful for forces in the range from 1 mg to 50 g (Brandt et al., 1976). They approach isometric transducers although some movement is necessary to develop the strain. However, for extremely delicate movements, or movements where a tissue is sensitive to pressure caused by the measuring device, this range is still not sensitive enough. To solve these problems three other methods are available which measure movements nearly isotonically or indirectly by other means which do not require attachment to the tissue at all.

c. Capacitance

A transducer circuit was devised by John James to measure small delicate movements by the American cockroach ventral diaphragm (Miller and James, 1976). The circuitry employed an oscillator which was capacitor-coupled to a detector (Figure 4-6). The capacitor was designed with one movable plate, mechanically connected to the ventral diaphragm. When the diaphragm moved, the plate distance changed and the capacitance dropped. A drop in capacitance decoupled the oscillator signal causing a decrease in the output voltage signal. The 47 μF tantalum capacitor was capable of transmitting long-lasting changes in diaphragm tension as long

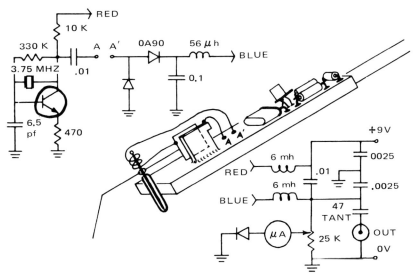

Figure 4-6. Capacitance transducer development to measure movements of the cockroach hyperneural muscle. The physical arrangement of the transducer is shown in the middle of the figure and the capacitor connects to the circuit at A A'. The left-hand circuit is mounted directly on the capacitor holder and connected to the right-hand circuit elements by cable. All capacitors are in mF unless specified. All diodes are the same and the meter is 100 μA connected to a 25-kΩ potentiometer. After Miller and James (1976).

as the output was connected to a high-impedance recorder (1 MΩ Brush 220 recorder for example). Upon activation, the transducer required 5 to 10 min to warm up to charge the output capacitor during which the output voltage drifted steadily and became constant thereafter at some value above zero. This circuit as used had no zeroing provision and the balance control on the Brush recorder was used to center the recorded signal at rest.

This transducer has also been used to record movements of the cockroach heart and can be applied with some changes to other smaller movements. Capacitance transduction has the advantage of extremes in sensitivity compared to strain gauges and potentially can approach perfectly isometric or isotonic measurements with further designing. The James device (Figure 4-6) has a nonlinear output because flat capacitor plates were used and capacitance edge-effects plus a dependence on the square of the separation difference produce a voltage per movement output which is greatest when the plates are closest together.

Normann (1972) constructed an isotonic transducer using capacitance. The blowfly heart preparation (Calliphora erythrocephala) was grounded and made to push on one capacitor plate. The stationary record plate was connected to a Clapp oscillator and a reactance converter. This was essentially similar to the capacitance transducer constructed from old a.m. radios (Miller and Metcalf, 1968). The latter device was highly susceptible to antenna loading and required careful screening to avoid stray capacitance pickup. Normann (1972) gave no details about his capacitance device, in this regard, but the John James device (Miller and James, 1976) is virtually free of interference.

Mr. R.M. Wise (Medical College of Virginia, Virginia Commonwealth University, Richmond, Virginia 23298) has developed a capacitance transducer with a linear output for use as a muscle transducer (Figure 4-7). The homemade device was modeled after one by Huxley and Simmons (1968) and Cambridge and Haines (1959). It has a sensitivity of 1.0 mV/mg and is linear from 0 to 100 mg. The transducer is capable of measuring millisecond transients of single muscle fibers with a maximum force of 60 to 80 mg and an accuracy of ± 3 mg. The natural resonant frequency of the transducer is 5 kHz, while that reported by Huxley and Simmons (1968) was over 10 kHz.

d. Inductance

Véró and Salánki (1969) published a detailed description of a movement detector based on inductance which is similar in principle to the John James device (Miller and James, 1976). The circuit was designed to measure the movement of mussel shells but may be adaptable as a trans-

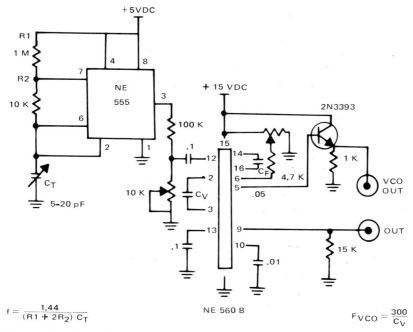

$$f = \frac{1.44}{(R1 + 2R_2)\, C_T}$$

NE 560 B

$$F_{VCO} = \frac{300}{C_V}$$

Figure 4-7. Capacitance transducer developed by R. M. Wise (1975) to measure muscle movement. The circuit uses NE 555 and NE 560B elements and 2N3393 transistor to produce a stable fast response without the use of tuned circuits.

ducer (Figure 4-8). The oscillator was designed for long life drawing 1 mA current. The arrangement was not susceptible to interference, but care had to be taken to ensure against the detector coils coming too close together as this caused the resonant circuits to detune. Any interruption at the detector transformer (Figure 4-8) could cause a drop in recorded signal. The original publication reported a gain equalizer stage inserted between the detector and recorder without explanation.

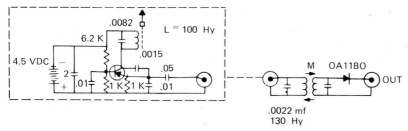

Figure 4-8. Véró and Salánki (1969) inductance movement detector. Once L was adjusted to the proper resonant frequency, the oscillator could be sealed and placed in water with the detector transformer connected to cable.

e. Impedance Conversion

Often a new development in one field of science has an unexpected application in others. Measurements using impedance are one example of this phenomenon. The principle itself is not new (Nyboer, 1959). Impedance bridge circuits were designed to measure changes in volume of body fluids of vertebrates (defined as impedance plethysmography) and have been used for some years.

Instrumentation has become available fairly recently (Weiss, 1973). Heinrich (1970, 1971) first employed a device termed an Impedance Converter (Model 2991, Biocom Instrument Co.) for recording the heartbeat of the moth, *Manduca sexta*. In the Biocom 2991 the output of a load-sensitive oscillator is impressed across a tissue and fluid. Changes in the tissue impedance modulate the amplitude of the oscillator signal which is converted to voltage changes at the output of the converter. Since the impedance of an organ and associated fluid is likely to be complex, it is difficult to calibrate the output voltage in any standard way. (For a short discussion of impedance design problems see Geddes, 1970.) Rather, in practice, amplitude of the output voltage is roughly correlated with the amplitude of contraction when the electrodes are stationary near moving tissue.

A variety of electrodes may be used with impedance conversion. We use large insulated copper wires and between uses cut off the tip ends which become corroded from being in contact with saline solution (T. Miller, 1973). Heinrich (1970) used silver and later 36 gauge copper (Heinrich, 1976) wires in his studies. The wires (200 μm) need only be near the tissue of interest; it is not necessary to touch the tissue. If the saline level falls below the exposed bare copper tips, then large voltage fluctuations are recorded which interfere with signals of interest. No such problem occurs with wires implanted in whole insects. One needs only to ensure that the wires are firmly fixed in place to eliminate movement artifact.

Even extremely small movements can be detected by impedance conversion, depending on the positioning of the recording electrodes. The physical layout of electrodes, leads, and recording instrument is not critical and little or no electrical interference is encountered. Thus, this principle is easily adapted to classroom work (cf. Part IV, 3, c).

Impedance converters may be used for recording tissue movements other than heartbeat. I have attempted only briefly to record writhing of single Malpighian tubules (from cockroach, *Periplaneta americana*) and abandoned the effort without much success. Movements of the ventral diaphragm of *Periplaneta americana* are very readily recorded with impedance converters; however, since the diaphragm is a flat structure, complex voltage responses are produced from simple movements, and we resorted to other methods for recording diaphragm movements (Miller and James, 1976).

f. Position Sensors

Position sensors were used by Delcomyn (1971, 1973) to detect leg movement and Sandeman (1968) and Barnes (J. Barnes, 1973, Zoology Department, University of Glasgow, Scotland, personal communication) to measure crab eye cup movement.   Also the Hall effect devices described in the previous section (Part III, 6, d) are a type of position sensor.

Sandeman (1968) used a fairly straightforward device for position sensing (Figure 4-9). The output from a sinewave oscillator was connected to a moving metal wand attached to the eye cup of a crab and to the base of a transistor (2N1302, Figure 4-9). The wand was positioned between two metal spheres or sensors which were connected to the input of a differential amplifier (Tektronix 122 was used).

When the wand moved between the sensors, an output signal was delivered to the collector of 2N1302 where it was compared to the original oscillator signal. The 2N1302 transistor acted as a gate and rectifier allowing amplitude modulation of the signal which was smoothed by the resistor and capacitor between collector and ground. Thus a DC signal was provided at the output which was proportional to the position of the wand.

Barnes (personal communication, 1973) developed a similar device (Figure 4-10) with a function like that of Sandeman (1968); however, more circuit detail is included. The preamplifier was mounted as close to the preparation as possible to avoid interference.

Marrelli and Hsiao (1976) have recently described a position sensor, which was designed for use under water, to detect the angle between segments of a crayfish limb. Two driving electrodes were constructed of

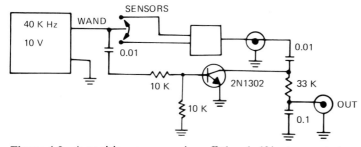

Figure 4-9. A position sensor using off-the-shelf instruments by adding a few circuit elements. The original description by Sandeman (1968) has been modified to include a resistor in the transistor base circuit (Sandeman, personal communication, 1977).

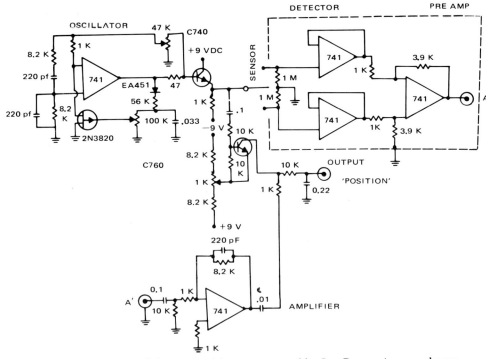

Figure 4-10. A more elaborate position sensor used by Jon Barnes (personal communication, 1973) and employing operational amplifiers. All capacitors in mF unless indicated otherwise. The output marked A is connected to A'.

0.25-mm spheres made by melting silver wire 70 μm in diameter. The wires leading to the spheres were insulated from the water with red-x Corona Dope (No. 50-2, G.C. Electronics). The bare spheres were epoxied to the proximal limb segment on either side of the joint pivot.

The sensing electrode was adjusted on the distal limb segment adjacent to the driving electrodes so that it moved in a plane containing all electrodes and perpendicular to the pivot of rotation of the limb joint. The driving electrodes were used as dipoles to develop a circular electric field which was centered on the pivot of the joint.

The output from a 100-kHz oscillator was impressed across the driving electrodes. As the sensing electrode rotated with respect to the driving electrodes, it sampled the electric field along the electric field lines. The sensor signal was amplified, detected, and then filtered to provide a DC output voltage. The sensing electrode traced an arc and its position was defined as the angle formed by a line (r) joining the midpoint between the driving electrodes and the sensing electrode and another line (d) joining both driving electrodes. The midpoint between the driving electrodes had to correspond with the pivot of rotation of the joint.

The best results were obtained if the distance of "d" was 3 mm and "r" was about 4.5 mm or $3d = 2r$. The DC output voltage was very nearly linear with respect to the angle formed by "r" and "d" from 0 to 180° providing that the electrodes were properly aligned. The 100 kHz driving signal was chosen so as not to interfere with ordinary nerve and muscle potential recorded extracellularly from the same limbs whose position was being detected. The circuit of this position sensing detector bears a superficial resemblance to the capacitance transducer which was designed by John James to measure small movements (Miller and James, 1976).

g. Phototransducers

A number of workers have used light beams to detect small movements. In fact, the light-sensitive diode or photoresistor or phototransistor may be one of the electronics devices most used in measuring movements of whole insects or parts of insects or insect organs. Several circuits using light beams have already been described above and any of these could be used in devising phototransducers.

The use of phototransducers to measure movement is limited not as much by the technology as by the experimenter's own ingenuity in constructing the apparatus. Noteworthy in the category of adaption are Peter Miller's successes in recording the movement of spiracle valves in cockroaches (Miller, 1973, 1974b). Small mirrors were attached to the valves of the left and right spiracles, and a beam of light was reflected off of the mirror onto a phototransistor nearby.

The circuitry for recording movements of the spiracles was the same as used by Miller for recording movements of spiracle valves in mantids and cockroach (Miller, 1969, 1971a) and was that used by Robert Pickard and Peter Mill for measuring ventilation movements of the abdomen in dragonfly larvae (Pickard and Mill, 1972, 1974, 1975). All of these studies employed the OCP 71 phototransistor in very simple measurement circuits (cf. Figure 2-6).

An isotonic transducer using a light path was described by B. J. Cook (USDA, Western Cotton Research Labs, 4135 East Broadway Road, Phoenix, Arizona 85040) for use in recording movements of isolated preparations of hindguts or foreguts from various insects (Figure 4-11). According to Benjamin Cook, the device was modified from a similar one used by Ernst Florey (Fachbereich Biologie, University of Konstanz, 775 Konstanz, Germany) to record the movement of Crayfish heartbeat.

The mechanical portion of this isotonic transducer includes the use of a semibalanced Palmer lever which interrupts a light path (cf. Davson, 1964, p. 914). The light source and the cadmium selenide photoconduc-

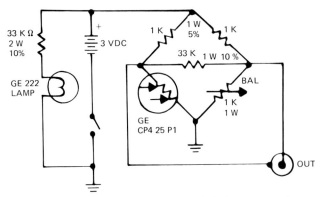

Figure 4-11. A circuit diagram for a light beam transducer. A vane attached to the moving tissue interrupts the light from the GE 222 lamp falling on the GE CP425P1 photoresistor. The balance control may be made more sensitive by use of a 10 turn potentiometer (Ben J. Cook, personal communication, 1976).

tive cell were connected to a common power supply and connected in a bridge circuit (Figure 4-11).

A warm-up period of 5 min was necessary to avoid drift of recorded DC potentials. The lever was adjusted to about the middle of its travel for maximum linearity in response. The balance or zero control was used to adjust a recorder pen to a standard position, and the mechanical lever could be balanced with added weight to suit the preparation.

There has been an increasing adaptation of optics in electronics. Optical scanners are now available which can detect stripes about 25.4 μm wide (Nano-skan® fiber optic skanner S2005-3 LED) (light-emitting diodes) and photodetectors. The light output and return path to the detector are housed in a fiber-optic (Skan-A-Matic Corp., Appendix) which can be aimed at an assembly line or an insect container.

The General Electric H13 A1 and A2 photon-coupled interrupter modules have an interruptible optical path. The H13 series offers either phototransistor output or photodarlington output. Other similar units have SCR output circuits which can switch a power line directly.

Another category of interrupter is the reflective coupler which is used to detect dark objects on a light field. In this case the response depends on the distance between the optical coupler and the object or surface being detected.

The RCA CA 3062 photodetector and power amplifier operates on 5 to 15 V and provides 100 mA output, enough to drive a relay or thyristor (Figure 4-12) (RCA publication No. SSD 210, *RCA Integrated Circuits*). Here phototransistors are used as the detecting elements, and detection and power amplification are accomplished in the same package.

An alternative circuit uses photodiodes followed by amplification (Dobkin, 1973) (Figure 4-13). Photoresistors have low linearity and low

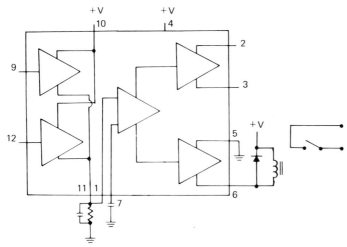

Figure 4-12. Circuit diagram of the RCA CA 3062 which combines a photocell and operational amplifier in one package. Suggested external connections are included for relay operation. From the RCA publication No. SSD 210, p. 172.

speed of response. Photodiodes have a low-level sensitivity but a better linearity than photoresistors or phototransistors. For extremely critical light measurements, the circuit shown (Figure 4-13) can resolve 1 nA from a photodiode which has a shunt resistance as low as 50 kΩ, and the device can detect light at 0.001 foot candles when using extremely high feedback resistances.

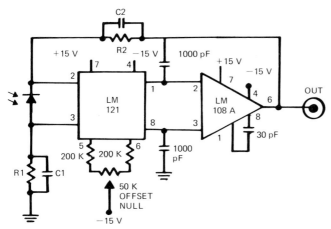

Figure 4-13. A circuit for use with a silicon photodiode plus amplification for highly sensitive light detection. The feedback circuit elements C1 and C2 and R1 and R2 are selected depending on diode capacitance and output voltage for maximum current. After Dobkin (1973, personal communication).

The feedback resistors R1 and R2 are selected by dividing the maximum output voltage, say 10 V, by the current flowing in the diode for a particular light intensity application. Thus if the maximum light intensity of interest from a dim source produces 1 μA in the photodiode, then the resistance of R1 and R2 would be half of 10 MΩ from Ohm's law. The capacitors C1 and C2 are each selected for about the value of the capacitance of the photodiode which is usually listed by the manufacturer.

h. Hall Effect

Hall effect devices were described in Part III, 6d for use with limb movements and in ventilation movements.

i. Contact Thermography

Thermistors are sold with various ratings including a dissipation constant, time constant, and standard resistance. For example, Fenwal thermistor GA 51J1 (Fenwal Electronics, Inc.) is a bead thermistor of 43 mil diameter (1.09 mm) rated at 100 kΩ at 25°C and decreases resistance by a factor of 10.3 between 0 and 50°C. The dissipation constant is 0.7 mW/°C. This is the amount of power in milliwatts that will raise the thermistor 1°C above its surroundings (Fenwal Thermistor Manual, No. EMC-5).

If the Fenwal GA 51J1 is connected across a 6 V DC battery at 25°C, then by Ohm's law the current flow would be 60 μA. Power, then, would be current times voltage or 0.36 mW. Under these conditions with a dissipation constant of 0.7 mW/°C, the thermistor would be heated about 0.5°C above the ambient 25°C.

The use of contact thermography is also described in Part IV, 3 where methods of recording are covered. Basically contact thermography is an anemometer principle depending upon heating a thermistor slightly above ambient temperature. Changes in flow of surrounding fluids cause changes in the dissipation of heat and therefore the resistance of the thermistor changes. The thermistor need not be heated more than 0.5°C above ambient body fluid. Then using a bridge circuit, 0.01°C temperature changes can be recorded and these minute fluctuations will reflect changes in fluid movement. Thus contact thermography is useful in detecting flow rates as well as the activity of pulsatile organs. Evidently, resistance changes in the thermistor are caused by changes in the rate of dissipation of heat. The heat dissipation is presumed to be higher for greater movement of fluids past the thermistor.

Wasserthal (1975, 1976) used contact thermography to record accessory pulsatile organ and heartbeat activity in Lepidoptera. Bernd Heinrich (1976) used thermocouples to measure temperature from the bumblebee abdomen. These temperatures varied at a rate which coincided with pulsations of the ventral diaphragm and therefore probably reflect pulses of higher temperatures of hemolymph reaching the abdomen from the warmer thorax. This points out one problem with contact thermography: one cannot be certain of the absolute temperature of the fluid in contact with the thermistor. A fast pulse of warm fluid may cause the same change in thermistor resistance as a slow pulse of cool fluid.

One tremendous advantage of contact thermography is in the longevity of implanted thermistors. No corrosion of metal surfaces occurs, so recordings are possible throughout a life stage. Wasserthal (1976) recorded the heartbeat of Saturniid silk moths and Brassolidae butterflies through pharate and adult stages without anesthesia and without apparent damage to the insects.

j. Producing Movements

Opposed to measuring movements is producing movements. This normally is encountered in dealing with sensory responses of chordotonal organs or proprioceptors or other mechanoreceptors.

The most versatile movements are obtained from electromagnetic devices operated by signal generators (Vedel et al. 1971). An example of this was the setup used by Rice and Finlayson (1972) for flexing the anterior wall of the cibarial pump of adult blue blowflies, *Calliphora erythrocephala*, while recording from the labrocibarial sensory nerve. A Pye-Ling V47 electromagnet device (Collins Electric electromechanical chopper) was used to move single tarsal hairs on the metathoracic leg of adult grasshoppers, *Schistocerca gregaria, Romalea microptera*, and *Locusta migratoria* (Runion and Usherwood, 1968), during extracellular recording of sensory activity ascending the tibia. The legs were removed at the tibia–femur joint.

Hugh Spencer (1974, personal communication) also used an electromechanical device for moving a hair receptor on the trochanter of adult male American cockroaches. A miniature loudspeaker was stripped of the cone and metal surround. A mica disc was glued to the armature coil and a 26 gauge tungsten needle was glued to the center of the mica disc. The needle was waxed to the hair of interest while nerve impulses were recorded with a 25 μm diameter gold wire placed 0.5 mm from the hair through the cuticle. Only the isolated femur was mounted. Chapman and Pankhurst (1967) also moved sensors on the cockroach leg. They stimulated campaniform sensillae by moving the cuticle with a probe cemented to a lead zirconite piezoelectric bender made by Clevite 60404 PZT Bimorph (Chapman and Nichols, 1969).

Möhl (1972) fashioned a rather complex device to stretch the foregut wall of the cricket *Acheta domesticus* L. He connected a motor and gear box by means of flexible extension links to move one axis of a Narishige micromanipulator. A potentiometer connected to the same axis knob was used to monitor movements.

The Zoology Department at the University of Glasgow, Scotland, also used a similar device to monitor movements. In this case movements of the grasshopper leg are produced by a cam connected to a kymograph (or any other turning motor) (Figure 4-14a).

Locusts (or grasshoppers) are fixed to a Plexiglas sheet with strips of plasticine. The femur of one hind leg is secured in line with a rod so that the tibia is free to move. Any misalignment between the rod and femur would put strain on the femur–tibia joint and produce complicating sensory impulses. To eliminate extraneous sensory impulses, the tarsus is removed by cutting at the distal end of the tibia and the crural nerve supplying the leg from the metathoracic ganglion must be cut as well. To cut the crural nerve, it is necessary to make an incision in the ventral thorax as explained elsewhere (Hoyle, 1968). Another alternative would be to use only the whole isolated leg severed from the body, as Spencer did.

A pin is attached to the rod and pushed through the stump of the tibia (Figure 4-14) taking care not to strain the femur–tibia joint. The rod is connected by levers and rods to the shaft of a potentiometer (POT.) which rides on the surface of a cam wheel. The cam may be driven by a kymograph motor or any synchronous motor, and the cam may be cut from any convenient pattern depending on the desired movement.

For recording nerve impulses, holes are punched through the cuticle of the femur about as shown (Figure 4-14b). Two wire electrodes of small diameter insulated copper are trimmed at the ends to provide a clean surface and then adjusted through the cuticle into the femur. Once inserted, an ordinary AC high-gain preamplifier may be connected to the electrodes and turned on. While observing the recorded voltages and preferably listening with an audio monitor as well, the tibia may be moved back and forth and the position of the electrodes adjusted until nerve impulses are recorded which coincide with the movement. This manipulation is necessary to bring at least one of the wire electrodes near the main leg nerve where it will pick up electrical activity from the sensory axons. It is important that, in locusts at least, the distal electrode not be over two-thirds of the distance from the coxa to the tibia since the chordotonal organ occupies the entire distal third of the femur (Usherwood et al., 1968).

Nagai (1970) used a slightly different electromechanical device to stretch the rectal longitudinal muscle of *Periplaneta americana* (Figure 4-15). The device was modified from the driver of a timer–marker pen from a kymograph and was essentially a relay movement. Since the electromagnet could be activated only as an all-or-none movement, the movement device was used for quick-stretch and quick-release studies only. Note that the total amount of movement depended entirely on the mechanical advantage of the distances along the steel wire as measured from the pivot point (Figure 4-15).

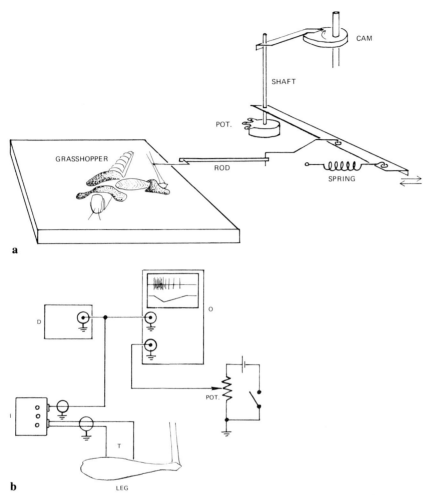

Figure 4-14. (a) Layout for the movement of a locust tibia. The cam is motor driven and pushes on a rigid set of levers which turn the shaft of a potentiometer (P) and move the tibia in a vertical plane. The potentiometer (POT.) is connected as shown on the circuit in (b) to provide a position-dependent voltage. (b) shows the circuit connections to record sensory nervous impulses ascending the crural nerve of the locust leg during movements of the tibia. I, amplifier; D, audio amplifier and speaker; O, oscilloscope; T, wire electrodes; POT., potentiometer.

Possibly the simplest type of sinusoidal movement device requires no electronics (Miller, 1968) but instead may be constructed from a mechanical metronome (Figure 4-16). If ordinary cotton string is wound around the shaft of the metronome movement and then through pulleys to the preparation, any small amplitude of movement may be obtained and the frequency is determined by the frequency chosen for the metronome. The resulting movement resembles the motion of a pendulum and is not

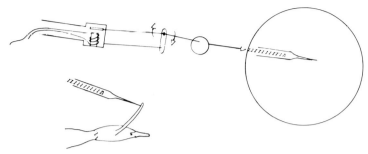

Figure 4-15. An electromagnet arranged in a holder as a simple device to produce movements. Completing the electromagnet circuit to a voltage source causes the probe to pivot around a fixed fulcrum (curved arrows). The tip of the probe is shown in greater detail to the right and is attached to the tissue (inset). After Toshio Nagai (personal communication, 1975).

truly sinusoidal. In the example shown the metronome movement device was used to lift the ventral part of a single heart chamber in a dorsal-side-down semiisolated preparation. The movement was accomplished by pulling on a silver rod placed in the lumen of the chamber.

Movements are produced in tubular organs by inserting a cannula and introducing pressure, either hydrostatic or air. Gelperin (1967) accomplished this with the foregut of the blowfly *Phormia regina* by using hydraulic pressure. Stretch receptors were demonstrated by recording nerve impulses in the recurrent nerve during foregut stretching. Ned Smith (personal communication, 1968) accomplished this with the heart of *Periplaneta americana* in a similar manner and recorded sensory nerve activity from the lateral cardiac nerve cords.

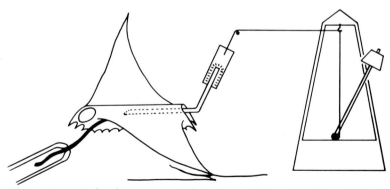

Figure 4-16. A simple movement detector made by attaching a string to the shaft of a metronome and to a pin which was inserted into the heart of American cockroach through a transverse section of the dorsal vessel. After T. Miller (1968).

k. Pressure Transduction

Sláma (1976) recently described a technique for recording insects' hemolymph pressure over prolonged periods of time. The method includes implantation of a disposable needle into insects *(Sarcophaga scoparia* larvae and *Pyrrhocoris apterus* adults were used). The needle transmits hemolymph pressure changes via saline solution and appropriate tubing to a tensiometer which records the pressure changes. A series of valves and reservoirs are included for continuous calibration, temperature and pressure compensation, and zero adjustment.

3. Organ Preparations

a. Introduction

The historical development of techniques used for the recording of movement or electrical activity in insect organs has been similar to the history of improvements in actographs, at least in the technical aspects.

Mechanical devices were used for actographs at first with kymographs and smoked paper. About 1930, several groups began using the mechanical interruption of light beams to measure small movements. With postwar advances in the electronics industry, devices based on electrical principles and amplification were used more frequently; however, simple force transducers and strain gauges, while well developed for vertebrates, did not find application in insect work. This was primarily due to a lack of ability to measure forces in the low and submilligram ranges.

With modern technology, there are a variety of instruments and principles available which are useful in measuring the activity of insect organs. Traditional laboratory exercises in insect physiology have included the use of microscope and stop watch to measure rates of rhythmic contractions of heartbeats for example. Newer methods replace the stopwatch with accurate and precise chart records of heartbeat movements. These records contain far more information than the older observations and enable the researcher–teacher to bring a fair amount of sophistication to the classroom. The same is true for any other organ movement.

b. Ventilatory Movements

Dragonfly larvae have a unique rectal chamber which undergoes tidal ventilation. The first extensive mechanical records of larvae ventilation movements were by Wallengren (1914).

More recently Pickard and Mill (1974) have examined ventilatory movements of *Anax imperator, Aeschna cyanea,* and *Aeschna juncea.* A Bolex 16-mm recording camera was used to record abdominal movements by photographing unrestrained dragonfly nymphs at 18 to 64 frames/sec in pond water in a glass container. Dorsoventral contractile forces were recorded by means of strain gauge transducers, and sternal movement was recorded with phototransistors (Pickard and Mill, 1972) (cf. Figure 2-6). In addition, expiratory water flow was recorded using a flow vane (Mill and Pickard, 1972).

Figure 4-17 shows the Perspex chamber used for measuring abdominal movements of dragonfly nymphs. The body and limbs of larvae were restrained ventral side up by clamps of bent wire or by a sheet of polythene. The caudal abdomen was fitted snugly through a slot in a central partition and sealed there by applying a small amount of ordinary vacuum grease which almost entirely prevented water leaks. The larvae lay in a formed depression in the wax portion of the chamber. Pickard and Mill (1976, personal communications) both report some occasional struggling by the larvae, but most of the time the larvae were restful.

Forces exerted by one of the dorsoventral muscles used for respiration were measured by an SRI isotonic strain gauge (see Mill and Pickard, 1972, for a diagram of the muscles of the dragonfly larvae showing the position in the abdomen). To attach the strain gauge, the insertion of a single dorsoventral muscle was located on the pleural cuticle of an abdominal segment.

Two holes 1 mm apart and about 70 μm in diameter were drilled

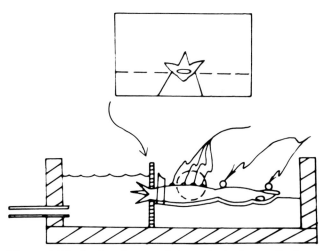

Figure 4-17. A chamber used to restrain dragonfly larvae and to measure abdominal movements. The dashed circle shows the position of a light beam transducer in cross-section. The inset shows details of the partition as viewed from the water chamber toward the wax chamber. The dashed line represents the wax level. Restraining clamps were placed on the head and thorax to reduce spurious movements (double arrows). Redrawn from Pickard and Mill (1972).

through the cuticle near the posterior margin of the insertion of the dorsoventral muscle. Either a spade drill (obtained from a wholesale jeweler) or a pin was used to make the holes. Then a 48 SWG wire was threaded in one hole and out the other; the exposed ends of the wire were twisted and attached to the strain gauge.

The area of the cuticle including the muscle insertion and the threaded wire and strain gauge attachment was cut away from the pleuron so that the full force of the dorsoventral muscle was recorded on the strain gauge.

The muscle potentials of respiratory musculature were recorded with wire electrodes using procedures similar to those described in the preceding section. Forty-eight SWG Lewcosal-covered copper wires (obtained from the London Electric Wire Co. and Smiths Ltd., Lewcos House, Lwr. Wortley Road, Leeds, England) were inserted through prepunched or drilled holes in the cuticle and into or adjacent to specific muscles. Within 1–4 hr a seal was formed around the wire by the wound-repairing properties of the hemolymph. A bit of Eastman 910 adhesive was placed on the cuticle and wire electrode to avoid movement artifact and to prevent damage.

Either one electrode was implanted in a muscle with an indifferent electrode in the body cavity or two electrodes were implanted in the same muscle. Muscle potentials were recorded with Grass P-4 preamplifiers.

To record ventilatory and spiracular movements in mantids, *Sphodromantis lineola*, P.L. Miller (1971a) strapped adults ventrum up to a Plasticine block. A thread was waxed to an appropriate sclerite and attached to the RCA 5734 transducer (for appropriate circuitry see Part IV, 2).

Movement of spiracles was transduced by placing a small (0.1 mg weight) mirror on a spiracle valve pretreated with a small amount of petroleum jelly. A light beam was reflected off the mirror onto a phototransistor (Mullard OCP 71, cf. Figure 2-6 for an appropriate circuit). The phototransducer was arranged so that only the response of the valve opening produced appreciable beam deflection; movement during ventilation or whole body movement did not alter the beam path enough to change the recording appreciably.

P.L. Miller (1974a, b) reported the measurement of ventilatory airstreams in the tracheal system of the mantid, *Sphodromantis,* by use of a thermistor inserted into a prothoracic trachea. Since this thermistor was heated 1°C above ambient, principles of contact thermography were evidently used to determine the flow of air in the trachea.

c. Pulsatile Organs

There are several recent reviews on the insect heart. Chapters 3 and 4 of the 1974 second edition of *The Physiology of Insecta,* edited by M.

Rockstein, cover physiology and some recording techniques. Klaus Richter published a massive review on structure and function of invertebrate heart (Richter, 1973) with special attention to work published after 1945. David and Rougier (1972) recently reviewed culture of insect dorsal vessel. Some early techniques for studying the insect heart were described in a review in Chapter 10 of Insect Muscle (T. Miller, 1974b).

There are now several methods for recording the insect heartbeat. Each method imparts its own peculiar type of information. The descriptions below will start with the most recent technique and include many of the methods which have been used through the years.

Contact Thermography

Wasserthal (1975, 1976) recently published recordings of moth heartbeat and the activity of accessory pulsatile organs using "contact thermography." The method uses a thermistor bead mounted externally on the cuticle over a pulsatile organ.

When connected as shown in Figure 4-18, a steady low DC current flows through the thermistor raising it 1–2°C above ambient. The voltage output of the circuit shown falls off 50% below a heartbeat rate of 15 beats/min. Pulsations of the hemolymph on the opposite side of the cuticle cause slight changes in the convection of heat away from the thermistor. These changes in heat dissipation cause fluctuations in the resistance of the thermistor at the rate at which the adjacent hemolymph is pulsed by pulsatile organs. In effect, contact thermography is very similar to

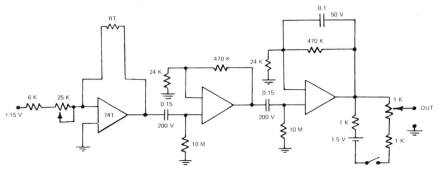

Figure 4-18. A simple circuit used for contact thermography including VECO thermistor 31A36. The variable 25-kΩ potentiometer controls the current through the probe thermistor RT. A power supply of ±15-V DC and zero is needed for the OP amps equivalent to Power Mate Model MD. All capacitors are in mF. All resistors are 0.5 W. A DC offset voltage is supplied by a pen lite battery at the output circuit (Salgado, 1976, unpublished). The OP amps are LF 356N.

anemometry which can employ thermistors to record air flow (Hamilton, 1976). When the overall temperature of the hemolymph fluctuates, this too causes changes in the resistance of the thermistors. Thus voltage records from contact thermography are typically pulsations modulated by slower changes (Figure 4-18).

Contact thermograph records of ordinary heart pulsations which send hemolymph in an anterior direction are readily distinguishable from records obtained upon reversal in direction of pumping of the hemolymph. Commonly referred to as heartbeat reversal, this phenomenon is characteristic of the heartbeat of many insects. Thus contact thermographs may reflect the actual net transfer of hemolymph besides providing information on temperature and heartbeat. This method can, therefore, offer more insight into the function of the circulatory system than many other methods.

Impedance Conversion

The impedance converter has not realized its full potential as applied to measurements in insect physiology. The use of impedance conversion in the classroom is especially appropriate because of its simplicity (T. Miller, 1973). The only drawback is a lack of ability to calibrate the magnitude of responses in units which are more familiar than ohms. This drawback is more than compensated for by the versatility of the impedance converter and especially the lack of interference one encounters in its use.

As described by Heinrich (1971) the magnitude of the impedance signal is roughly correlated with the amplitude of contractions. Figure 4-19 shows two records of the heartbeat of the third abdominal chamber of a semiisolated cockroach heart (*Periplaneta americana*) recorded by impedance conversion (Figure 4-19, i) and simultaneously by capacitance transduction (Miller and James, 1976, Figure 4-19, ii). When examining the record in Figure 4-19a, it can be seen that the greatest change in impedance occurs during the contraction of the heart, but that the negative peak of impedance (farthest excursion "down") does not coincide with maximal contraction (farthest excursion "up" by capacitance transducer). The physical position of the electrodes is shown in the three-quarter view inset (Figure 4-19, inset).

Despite a lack of correlation between mechanical movement and impedance conversion, the amplitude of impedance is not similar to, but is roughly correlated with amplitude of movement (Figure 4-19b). Note in particular that impedance changes may be far more complex than mechanical movement (Figure 4-19b). Also the impedance measurement tends to show an exaggeration of slight differences in amplitude of contractions.

The latter phenomenon depends a great deal on the exact position of

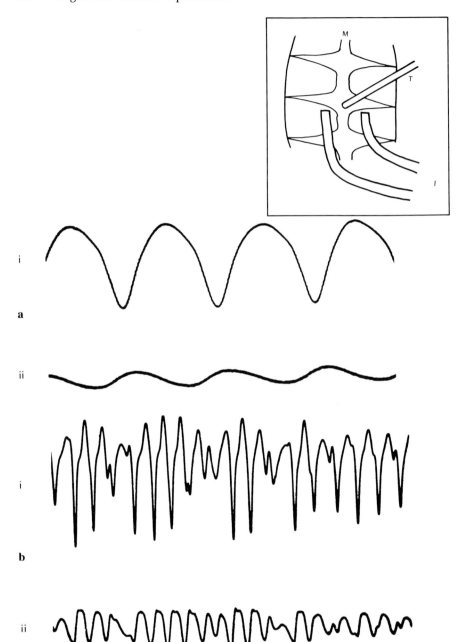

Figure 4-19. A comparison between recordings of cockroach heartbeat by impedance conversion (traces i in a and b) and by a capacitance movement detector or
isotonic transducer (traces ii in a and b). The inset shows the position of impedance electrodes (I) and transducer probe (T) which rested on the surface of the
dorsalside-down heart preparation (M).

the electrodes, but, since slight changes are so obvious, even the subtle actions of drugs are evident with impedance conversion measurements. Another method of bringing out subtle differences in individual heartbeats is to measure the instantaneous frequency (or the *period* between beats) of each heartbeat (Normann, 1972; Collins and Miller, 1977). This can be done with a rate meter such as the one produced by Ortec (Model 4672, instantaneous time/frequency meter) or a tachometer (Normann, 1972). When connected as shown in the inset in Figure 4-20, a heartbeat record is obtained and displayed on a dual-channel recorder along with an impedance converter record of the heartbeat (or any other transducer may be used). A capacitor was placed across the Biocom 2991 output circuit and selected to filter out high-frequency signals of very small amplitude which sometimes are not discernible at the impedance converter output but which are counted by the Ortec 4672. Many rate meters do not have ranges low enough to record heartbeat rates. A full scale range of at least 5 Hz is needed for most insect heartbeat rates. This is 300 beats/min and some hearts can beat faster than this, although most are within this range, especially Orthopterans.

As seen in the recordings of Figure 4-20a, the heartbeat record has some amplitude variation which is seen as steps in the rate record (lower trace). After application of one 35 μl drop of 10^{-3} *M* aminophylline (at the arrow) the amplitude became shorter and the average frequency increased; however, both the heartbeat record (top trace) and the rate record (bottom trace) show a marked irregularity in heartbeat.

The heartbeat changes caused by aminophylline are contrasted to the response to one 35 μl drop of 7.5×10^{-6} *M* 5-hydroxytryptamine (serotonin) under similar recording conditions, using another abdominal heart preparation (Figure 4-20b). In this case, however, both heartbeat and rate records show a slight increase in average heartbeat rate, a decrease in amplitude, and a discernible development of uniformity in the heartbeat record. The uniformity of heartbeat rate can best be appreciated by comparing the upper traces in Figure 4-20a and b between the dots.

Optical Method

Interference of a light path has been a traditional method of recording heartbeat in insects. Tachibana and Nagashima (1957) and Biston and Sillans (1976) mounted a light source and photocell near larvae of Lepidoptera. Phaneuf et al. (1973) described a simple method for recording the heartbeat (or any movement) by using a binocular microscope (Figure 4-21). A light source is fitted over one eye piece and a cadmium sulfide photoresistor is fitted over the second eye piece. When properly focused, the optical method can record heartbeat without connections of any kind to the insect.

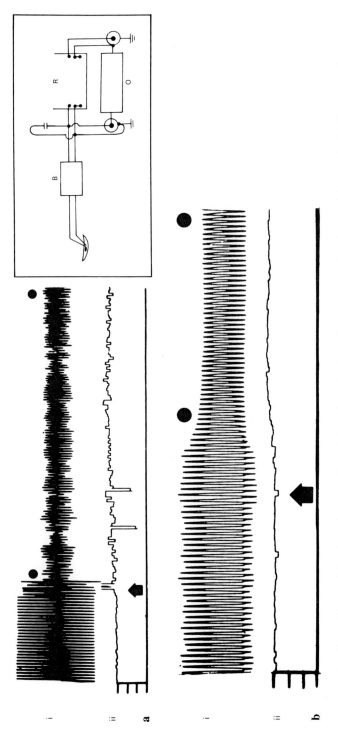

Figure 4-20. Recordings of the American cockroach heartbeat by impedance conversion (traces i in a and b). The traces in a and b are integrated and the rate is applied to the second channel of the monitoring recorder (traces ii). (a) shows development of an irregular heartbeat upon application of one drop of 10^{-3} M aminophylline (arrow) with immediate response. (b) shows smooth heartbeat response to the application of one drop of 7.5×10^{-6} M serotonin to the semiisolated heart preparation. The inset shows the connection of leads to the Biocom impedance converter, B, and the Ortec 4672 rate meter, O, and Brush 220 recorder, R.

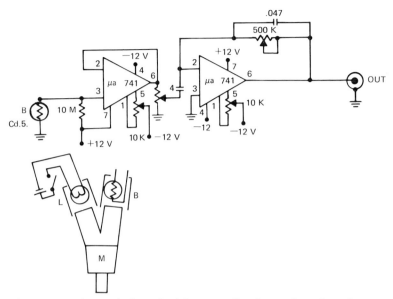

Figure 4-21. An optical method for recording insect heartbeat from unrestrained insects.   The inset shows lamp, L, and photoresistor, B, mounted on the eyepieces of a dissecting microscope, M.   From Phaneuf et al. (1973).

The obvious drawback to the optical method is movement by the subject. If the insect moves, all calibration and measurement cease. If the insect is restrained, this imposes other restrictions. The great advantage of the optical method is its simplicity.

Williams et al. (1968) carried the optical method to extremes when they reflected a beam of laser light off the blowfly heart. The Williams device is really a variation of the older light beam and mirror devices (Yeager, 1938; Uramoto, 1932; Duwez, 1938; Crescitelli and Jahn, 1938; Hamilton, 1939; de Wilde, 1947; Kooistra, 1950; Jahn et al., 1937).

Electrical recordings

Normann (1972), Moreau and Lavenseau (1975), Crescitelli and Jahn (1938), Takahashi (1934), Tenney (1953), Richter (1967), Duwez (1936), Irisawa et al. (1956), Quennic and Campan (1972, 1975), Campan (1972), and Tachibana and Nagashima (1957) have recorded "electrocardiograms" or electrical potentials accompanying contractile activity in insect dorsal vessel. McCann (1969) and Katalin Salanki-Rózsa (S.-Rózsa and Véró, 1971) have recorded similar extracellular potentials from the myocardium of moth and Orthoptera respectively using suction electrodes instead of extracellular electrodes.

Extracellular recording of this nature is susceptible to electrical interference, largely because the band width of measuring instruments allows extraneous signals through the amplifiers. Ground loops so common to ordinary electrophysiology are also possible. A wide variety of electrical wires have been used for electrodes, and the connection between paired electrodes and measuring amplifiers is similar to extracellular nerve recording. Most workers use amplifiers similar to those described in Part I, 8, b (high-gain amplifiers).

American Cockroach Heart Procedures

Plexiglas chambers for maintaining insect heart in semiisolated preparations are naturally a matter of individual taste and depend on the experiment performed. Details are given of setups and procedures with the American cockroach, *Periplaneta americana,* so that they may be copied or improved upon.

Saline

Saline solutions used for insects are not in the least standardized. Amino acid constituents of hemolymph have been measured for a large number of species (Florkin and Jeuniaux, 1974) and although it is evident that amino acid levels are regulated (Collett, 1976) there is lingering controversy over the measurement of glutamic acid levels (Irving et al. 1976).

Many different saline solutions have been developed for use with the American cockroach alone, *Periplaneta americana.* Some of the more recent saline solutions are given in Table 4-1. Blood potassium when measured by flame photometry yields a value of 24.2 mM; however, when measured by potassium-selective electrodes, the effective concentration in the blood was 10 to 15 mM K although the values for K given by Heit et al. (1973) using spectroscopy were in the range 5–8 mM.

For years the standard saline for cockroach central nervous studies was Narahashi saline (Yamasaki and Narahashi, 1959) (Table 4-1). Narahashi saline was arrived at by using a larger sodium concentration (214 mM) than previous salines to coincide more nearly with the osmotic pressure of *Periplaneta* hemolymph. Hemolymph osmotic pressure was measured as equivalent to 224 mM NaCl by Ludwig et al. (1957), and Narahashi found that desheathed *Periplaneta* ventral nerve cords swelled in saline with a sodium concentration of less than 214 mM, other salts being the same.

It is thought that the sodium concentration of hemolymph is considerably lower than 214 mM and that the potassium concentration of hemolymph is considerably higher than 3 mM used in Narahashi saline. Evans (1975) reported that the Narahashi saline resulted in weight losses in intact (not desheathed) and ligated abdominal nerve cords of *Periplaneta*. To correct for the weight loss, the amount of sodium was decreased and potassium increased to correspond more closely to hemolymph values reported by Pichon (1970) (Table 4-1). The osmotic pressure was balanced initially by the addition of 20 mM glucose, but later this was left out since glucose was reported to interfere with the uptake of amino acids.

In his analyses of hemolymph ions, Pichon (1970) found wide variations in levels of ions even when taken from the same cockroach. He concluded that such wide variations occur normally on either side of genetically fixed mean values. Brady (1967a) had previously reported some difficulty in obtaining true values for potassium since blood cells sequestered unknown amounts and the blood cells were difficult to separate from plasma for plasma analyses. Potassium values from flame photometric analysis were found to be about double those obtained by using K-selective electrodes (Thomas and Treherne, 1975), and this supports the suggestion that the effective K concentration is somewhat lower than the total K content of hemolymph.

As one final note on cockroach salines, Crowder and Shankland (1972a) found that the early heart saline of Meyers and Miller (1969) was useless in maintaining active writhing of the isolated Malpighian tubules of *Periplaneta* (see Table 3 of Miller, 1975). The original heart saline was a copy on one developed by Yeager (1939) and Ludwig et al. (1957). Yeager reported an "improved" heartbeat upon the addition of magnesium (1 mM) compared to the same saline without magnesium. The saline was based on osmotic pressure of hemolymph and the ionic composition giving the longest heartbeat in isolation. The Ludwig and Yeager values for Na, K, and Ca are all uniformly higher than the values given by Evans (1975) and the latter are probably much closer to actual hemolymph values.

Crowder found that the Meyers and Miller (1969) saline could not support writhing of isolated Malpighian tubules of *Periplaneta* for more than 5 min. Dropping the Ca to more normal values did not improve the ability to maintain writhing, but halving the K increased writhing time to over 2 hr and dropping the K and Ca to near the values of Narahashi's saline increased writhing time from 6 to 12 hr.

More recent cockroach heart saline values (Table 4-1, Miller and James, 1976) resemble values arrived at independently by Evans (1975) with the exception of lower sodium values used by the latter. Both heart and Malpighian tubules do not appear to require a specific Na concentration in their salines, but they are sensitive to K and Ca and Mg (heart at least) levels.

In concluding these notes on cockroach saline solutions, a few added remarks may be helpful. Yeager (1939) noted that the addition of Mg ion

Table 4-1. Saline Solutions Used for Organ Preparations with *Periplaneta americana*[a]

Name of user and date	NaCl	KCl	CaCl₂	MgCl₂	H₂PO₄	HPO₄	HCO₃	Other
				(mM)				
Glaser (1925)	154.00	5.63	2.25				2.38	Glucose (13.88)
Eastham (1925)	128.33							Peptone (2 g/liter)
Hobson (1928)	157.76	2.95	1.98					
Levy (1928)(see Burton, 1975)	154.00	9.52	4.14		0.08		2.02	
Ten Cate (1924)	162.6							
Bělař (1929)	154.00	2.68	1.80				2.38	
Yeager and Hager (1934)	168.03	10.33	4.51		0.08		2.14	Glucose (5.55)
Yeager and Gahan (1937)	168.03	10.33	4.51					Dextrose
Pringle (1938)	154.00	2.68	1.80					
Yeager (1939)a	201.56	12.34	5.95					
Yeager (1939)b	187.02	21.06	7.66	1.06				
Griffiths and Tauber (1943)	250.33	6.04	4.51				2.26	
Krijgsman (1950)	168.03	10.33	4.51					Dextrose
Kooistra (1950)	154.00	9.50	4.13					
Ludwig et al. (1957)	188.22	18.78	9.91					
Yamasaki and Narahashi (1959)	210.2	3.0	1.8		1.8	0.2		
Schofield and Treherne (1978)	214.0	3.1	1.8		0.2	1.8		

							Other constituents
Treherne (1961)	155.02	12.29	4.51	3.93			Sugar, amino acids, buffered
Ting and Brooks (1965)	188.22	18.78	9.91				Sugars, yeast extract, vitamins, organic acids, amino acids, buffered
Brown (1965)	154.00	2.68	1.80				Glucose (22.2)
Yeager and Ludwig	188.22	20.12	9.01	0.98	1.67	1.43	Glucose (71.44)
Meyers and Miller (1969)	188.00	20.1	9.0				$MgSO_4$ (1.0)
Wareham et al. (1973, 1974a,b)	140.0	16.0	10.0	6.0		16.0	
Evans (1975)	150.0	10.0	1.8	0.2			
Bennett et al. (1975)	113.7	25.0	2.0		1.8		
Miller and James (1976)	200.17	10.73	3.40	0.01		2.14	
Pichon (1970)	119 ± 28	28 ± 20	3.1 ± 1.1	(hemolymph analysis)			$MgSO_4$ (0.996)

[a]See also Burton, 1975.

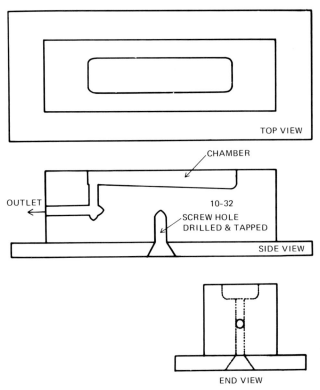

Figure 4-22. Construction details of a model perfusion chamber. Note the reversed slope of the floor of the chamber which allows saline solution to pool over the tissues then spill over into the outlet tubing.

in saline solutions resulted in a more regular heartbeat. We now feel that this was likely caused by an antagonism of Mg ion with Ca ion at the numerous cardiac neuromuscular junctions which decreased postsynaptic potentials and therefore reduced the effect of nervous activity from the cardiac neurons. This would result in a slightly greater expression of the myogenic heartbeat and would therefore appear to produce a more uniform heartbeat.

The membrane potentials of cockroach myocardium and Malpighian tubule muscles are at least partly dependent upon potassium ion gradients (Miller, 1975). Thus a saline solution should contain a K concentration, similar to that in the hemolymph.

Perfusion System

Our perfusion chamber consists basically of a Plexiglas block about 2.5 × 2.5 × 8 cm (1 × 1 × 3 in.) shown in Figure 4-22. A hole is drilled (No. 7 drill, 0.51 cm diameter) in the center of one end about

1.5 cm deep and tapped (1/4-20) for a stainless-steel adapter (#107A20, Chromatronix, Inc.). A trough is cut in the center of the top surface about 4 or 5 cm long, 5 mm deep, and 12 mm wide. Usually the trough is milled by 0.5 in. diameter (1.27 cm diameter) helix end mill to give a rounded bottom. The trough can be polished smooth so that it is transparent if transmitted light is used for bottom illumination. For the same reason, the perfusion chamber may be cut from a piece of 1 in. thick stock. This leaves the top and bottom surface transparent.

The trough may be cut flat and the bottom covered with Sylgard® resin which is transparent and accepts pins for securing tissue. The trough may also be cut at a slant for saline to form a pool and spill over into the outlet channel. The only problem with spill-over chambers is the hydrophobic nature of clean plastic which causes formation of drops. A periodic surge in fluid flow which causes dripping anywhere near the perfusion chamber disturbs delicate preparations. Dripping has been eliminated in the past by rubbing detergent on the outlet surfaces to reduce the surface tension and help wet the plastic to ensure a smooth flow of saline from the trough into the outlet tubing.

Another outlet arrangement which eliminates interruption of flow provides an adjustable constant level in the perfusion chamber by a remote vacuum connection (Figure 4-23). The saline level in the perfusion chamber can be adjusted to any height by adjusting the solution leveler chamber. The tubing connecting the two chambers should be of

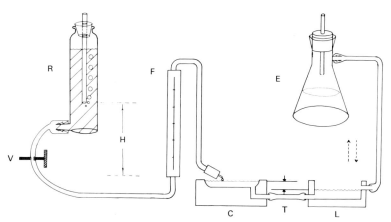

Figure 4-23. Perfusion system with a constant level output. The chamber (C) is the same as the one in Figure 4-22. A constant head of pressure (H) is maintained by the arrangement of the reservoir (R) as shown. A valve (V) and flow meter (F) deliver a constant amount of saline as long as the reservoir is kept above the top of the flow meter (if a bubble becomes trapped in the top of the flow meter tubing as shown, the flow would be changed). The exact level of saline in the chamber is determined by positioning the leveler chamber (L) (dashed arrows), as long as the connecting tubing (T) is of *large* diameter. The evacuation bottle (E) is connected to vacuum and normally held below the Plane of the leveler chamber.

large diameter and flexible. Food and medical grade silicone tubing is useful for this (Cole–Parmer Instruments, Inc.). Also the adapter connections, where metal is screwed into plastic, can be treated with RTV-108 silicone rubber adhesive (General Electric Co.) or equivalent to prevent leaks. A flow meter is helpful in maintaining a constant flow rate and the needle valve should allow precise adjustment of the flow rate.

For some perfusion systems, the tubing used for intravenous solution administration can be useful (I.V. Set, McGaw Laboratories, Division of American Hospital Supply Corp., Glendale, California 91201). These are sold with the fluid pathway sterilized and may be replaced when using drug solutions to prevent contamination of the perfusing system. The I.V. sets have standard Luer connections and ordinary laboratory supply houses provide a number of Luer adapters which make a variety of tubing configurations. We have used Popper and Sons Perfecktum®, Ace Glass, and Chromatronix for Luer adapters and Tuohy Borst adapters, three-way valves, and manifolds. Also Clay Adams P.E. 60, or P.E. 20 tubing is useful for transporting fluids with Tuohy adapter fittings and Gilmont provides flow meters.

When perfusion systems are used in drug studies, various compounds may be applied in a variety of ways. The system tubing may be interrupted at the manifold where compounds are introduced directly into the flow through a tube containing the drug. For long-term experiments, drugs may be applied in separate reservoir bottles. Drugs may also be injected into the saline stream immediately prior to delivery at the preparation independent of the flow system.

Denervation Procedure

The adult and nymph *Periplaneta* hearts contain paired lateral cardiac nerve cords which lie on either side of the dorsal vessel and accompany the vessel from the terminal abdominal segment up to the retrocerebral complex behind the brain (Alexandrowicz, 1926; McIndoo, 1939, 1945; Johnson, 1966). For some drug studies, the lateral cardiac nerve cords may be removed and the intrinsic myogenic activity of the heart recorded. In other studies, the nervous activity of the lateral cardiac nerve cords may be recorded.

The semiisolated heart preparation is usually the Type B preparation originally described by Yeager (1939) which includes the abdominal dorsum. It is advisable to dissect on the ventral side of the abdomen so as to include as much lateral spiracular tissue as possible in the final preparation. When the abdominal dorsum is mounted dorsum down, the heart tissues are exposed and accessible in this position.

In any case where lateral cardiac nerves are involved, the diagram in Figure 4-24 serves as a guide.

For location of operations on the lateral cardiac nerve cords, sharpened tungsten probes and fine forceps are helpful. The abdominal heart is suspended beneath the dorsal diaphragm (Figure 4-24e, Dd) and closely adheres to the dorsal cuticle. The lateral alary muscles (Figure 4-24e, Am) occupy the lateral edges of each dorsal diaphragm where the diaphragm is attached to the acrotergite in each segment. The position of this attachment and therefore the main point of suspension for the diaphragm is roughly opposite to a point midway between the paired ostial valves and the paired segmental vessels. Four pairs of segmental vessels are present in the second, third, fourth, and fifth heart chambers in the abdomen of *Periplaneta*. The walls of the vessels are extremely thin and usually transparent. They can usually be located where they form a channel in a perpendicular direction away from the midline of the dorsal vessel. The segmental vessels open immediately posterior to the main point of attachment of the dorsal diaphram, or, put another way, they end laterally immediately posterior to and beneath the lateral edges of the alary muscle "fans," as viewed in the dorsal-side-down preparation. These segmental vessels are not easy for undergraduate students to locate, but they serve as convenient landmarks.

To remove the lateral cardiac nerve cords in the fourth abdominal heart chamber, the dorsal diaphragm must be removed as it covers the pericardial sinus. Insert a probe just off the center part of the dorsal vessel and posterior to the ostial valves (Figure 4-24a). This area lies approximately midway between the lateral insertion points of the diaphragm for this chamber, and the tension in the diaphragm is in the direction indicated by the lines diverging from the insertion point toward the dorsal vessel.

The dorsal diaphragm is stretched and ordinarily under tension. If the probe is moved in the proper direction, the diaphragm will tear away from the midline of the dorsal vessel. If the probe is moved in the opposite direction, the dorsal vessel is likely to rupture leaving the heart split open along its ventral midline at the point of diaphragm attachment.

Once the diaphragm is torn away from over the area of the ostial valves (Figure 4-24b), the lateral cardiac nerve cords can be located by finding the ostial valves. The segmental nerve leaves the third abdominal ganglion and a small branch joins the lateral cardiac nerve cords at the level of the ostial valves. The final branch lies on *top* of the major longitudinal tracheal trunk (Figure 4-24, TR) near the heart (as viewed in the dorsal-side-down heart preparation).

The final connection of the segmental nerve is invariably accompanied by a small tracheole tube which usually makes a "Y" with the lateral cardiac nerve cord a small distance lateral to the ostial valve opening. Since the nerves are extremely difficult to see, it is usually simpler to find tracheoles.

With the dorsal diaphragm removed from over the ostial valves, a probe

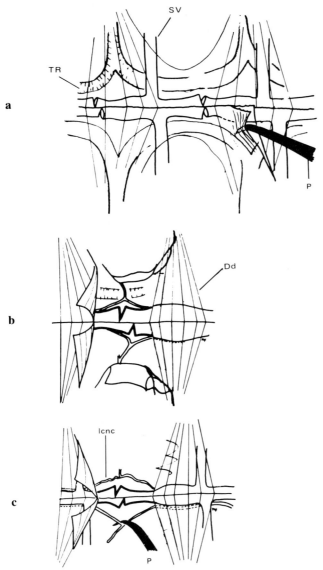

Figure 4-24. Drawings of the cockroach abdominal heart as it appears in the dorsum-down preparation. (a)–(d) shows operations to remove the lateral cardiac nerve cords (lcnc) using a sharpened tungsten probe (P) and forceps (F). The ostial valves are located first (Os) in between the segmental vessels (SV) in the middle abdomen. A probe is inserted just underneath the dorsal diaphragm (Dd) and torn anteriorly to expose the ostial area. The probe is inserted next to the heart tube at the ostial valves and moved laterally to tear away the lcnc. Then by pulling with forceps at one valve and teasing with a probe (d) the lcnc can be coaxed away from the heart. (e) shows a cross-section of the abdominal heart with appropriate structures for orientation. A recording electrode used with impedance conversion is shown (E) as well as the level of ordinary saline solution.

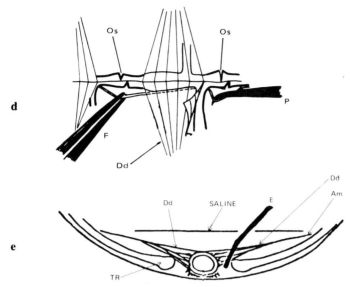

Figure 4-24 (continued).

is inserted next to the dorsal vessel at the ostial valve opening (Figure 4-24c). The ostial valve is the only place along the whole dorsal vessel where the lateral cardiac nerve cord leaves the surface of the heart, and therefore the probe must be inserted here to have a chance of tearing away the cardiac nerve cord. When the probe is gently moved laterally away from the heart, the lateral cardiac nerve cord can be seen as it pulls away from the heart surface with adhering tracheoles and pericardial cells.

The cardiac nerve cords are large in diameter near the ostial valves and taper anteriorly as they near the segmental vessels. The lateral cardiac nerve cords arborize near the base of the segmental vessels and send many branches laterally along the vessels. For this reason the nerve cord is more likely to break near the posterior base of the segmental vessel than at any other place along the dorsal vessel. To be assured of removing the entire segment of lateral cardiac nerve cord from one side of the heart, the cord must be held anteriorly and posteriorly and gently tugged back and forth simultaneously as shown in Figure 4-24d until it comes free from the area where the segmental vessel joins the heart.

If the dissection procedure is stopped at step c, the cardiac cord may be cut anterior to the ostial valve and the posterior stub pulled into a suction electrode. From this position, nervous impulses traveling up the cardiac nerve cord from cardiac neurons may be recorded. Since cardiac neurons are most readily found near the base of the segmental vessel, and since these neurons are stretch sensitive, Figure 4-24c represents a potentially useful preparation for the study of a stretch-sensitive neuron. Alternatively, the segmental nerve may be left intact and the neural coordination descending from the ventral nerve cord may be recorded (Miller and Usherwood, 1971).

Not all insect hearts contain lateral cardiac nerve cords. These are often found in the Orthoptera; however, the dissection scheme outlined here may be used as a guide to other Orthopteran hearts. The physiology of the cardiac neurons is largely unexplored at present and the valve cells at the opening of the segmental vessel are also well described (Miller, 1975) but unstudied beyond the dissertation work of Terry Beattie (cf. Miller, 1975). For example, the flow of hemolymph through the segmental vessels could conceivably be measured by contact thermography to study the control of the segmental vessel valve.

Thermistor thermometers can be purchased or constructed and used with perfusion chambers for documentation of precise temperatures. Recording room temperature as experimental procedure is outmoded when more accurate temperature measurements are now possible. Use of an ordinary incandescent lamp for tissue illumination can sometimes increase the temperature of a preparation markedly. Fiber optic illumination often eliminates heating from light sources.

The perfusion system shown in Figure 4-23 can be easily adapted to classroom measurement of Q_{10} for semiisolated heart or other rhythmically beating tissues when an impedance converter or other recording device is used. Impedance converter measurement would be compatible with the presence of temperature baths or other electrical applicances near the preparation.

Ventral Diaphragm

Insect ventral diaphragm studies include those by Richards (1963) who did a survey of insects, Dierichs (1972) on *Locusta migratoria*, Guthrie (1962) on *Schistocerca gregaria*, Shankland (1965) on the innervation of the diaphragm (called hyperneural muscle) in *Periplaneta americana*, Engelmann (1963) on the same structure in *Leucophaea maderae*, Miller and Adams (1974) on electrical properties of the muscle described by Shankland, and Miller and James (1976) on drug responses of the hyperneural muscle of *Periplaneta americana*.

Other work has touched on the ventral diaphragm of insects indirectly. Heinrich (1976) has recorded ventral diaphragm activity from queen bumblebee, *Bombus vosnesenskii*, by using impedance conversion.

Accessory Pulsatile Organs

Accessory pulsatile organs (APOs) which pump hemolymph to the appendages are well known in insects (Jones, 1977). Much of the information available on APOs is descriptive in nature. Kaufman and Davey

(1971) shocked the fused thoracic ganglion of *Triatoma* while observing the activity of the APO *in situ* in the tibia of a partially dissected preparation. Ten resting APOs were excited into activity by shocks delivered at 1 Hz.

Since amputation of the leg or severing the leg nerves resulted in cessation of pulsation of the APO in that leg, the APOs are evidently innervated. Dopamine increased the rate of pulsations at a threshold of $10^{-8}\ M$ while 5-hydroxytryptamine (5-HT) had inconsistent effects on APO activity.

d. Foregut

The foregut of insects has been the object of bioassay studies (Ten Cate, 1924; Kooistra, 1950; Hobson, 1928; Cook et al., 1969; Freeman, 1966; Beard, 1960; Brown, 1965), but it has been examined more for its role in feeding (Getting and Steinhardt, 1972; Gelperin, 1971; Graham-Smith, 1934; Engelmann, 1968; Möhl, 1972; Clarke and Grenville, 1960; Beard, 1960; Knight, 1962; Jones, 1960).

Stretch receptors on the foregut of the blowfly are of key importance in limiting feeding (Dethier and Gelperin, 1967; Gelperin, 1967). Stretch receptors on the foregut of the cricket interact with the motor neurons in the esophageal nerves to coordinate pulsations of the foregut musculature (Möhl, 1972). While it is not clear what role the cricket stretch receptors play in feeding, Engelmann (1968) hypothesized that for the cockroach, *Leucophaea maderae,* the degree of stretch of the crop and the consistency of the crop contents are registered by the stomatogastric motor centers which control crop emptying.

Besides demonstrations of stretch sensitivity of the foregut in insects, preparations have been developed to study the spontaneous activity of the foregut and to study the response of the foregut to various drugs.

Drug studies are often performed on isolated foregut preparations, and therefore the inherent activity of the foregut is of some importance. Yeager (1931) observed the crop of the cockroach, *Periplaneta fuliginosa,* while the intact insect was restrained between microscope slides. He concluded that the foregut was constantly active, but increased activity markedly immediately after a meal.

Yeager observed that the crop in the intact cockroach showed three movements which he classed as: (1) peristaltic contraction starting at the esophagus or pharynx and traveling posteriorly to the region just anterior to the gizzard, (2) antiperistaltic contractions (reversed peristalsis, Beard, 1960) which traveled interiorly at a rate somewhat faster than peristaltic waves, and (3) contractions of the whole posterior third of the crop. Reversed peristaltic waves sometimes originated after contraction of the posterior third of the crop and both of these types of contractions were associated with rhythmic contractions of the gizzard.

Contractions of the gizzard were variably affected by operations, but especially by decapitation. When excised through the thoracic body wall, the crop and gizzard were not active when immersed in 1% NaCl. However, when the head was circumcised and pulled away from the body with the crop and gizzard attached, the crop showed much activity while the gizzard contracted infrequently (Yeager, 1931).

The partially dissected gut of *Calliphora* (Graham-Smith, 1934) or *Phormia* (Knight, 1962) showed continuous nonsynchronous activity when left *in situ* in the body. When left *in situ* with every connection between the gut and the body severed except the esophagus, food was transferred from the crop duct to the ventriculus. Decapitation and removal of thoracic and stomatogastric nerve centers did not affect gut activity; however, complete isolation of the gut into saline or hemolymph caused loss of normal function. Completely excised guts were unable to transfer food from the crop to the ventriculus (Knight, 1962).

The crop of *Phormia* contracted forcefully and irregularly, but a moderately filled crop was far more active than an empty crop or a crop filled to distension (Knight, 1962). Yeager (1931) reported an increased activity in *Periplaneta fuliginosa* crops immediately after feeding. Brown (1965) half filled the isolated crop of *Periplaneta americana* before connecting to a perfusion system for drug studies. The completely excised gut of *Aedes aegypti* and other mosquito larvae (Jones, 1960; Schildmacher, 1950) showed constant activity.

Thus crop function appears "normal" when insects are opened only enough to expose the tissues of the gut, but left *in situ*. Beard (1960) found this for *Galleria mellonella* larvae and he noted that further dissection disrupted or modified foregut activity. Knight noted a difference between *in situ* activity of *Phormia regina* gut and activity of the completely excised gut. And Yeager (1931) noted the extreme sensitivity of gizzard movements to various ablations, dissections, and manipulations.

Beard Preparation

A unique preparation or procedure in the study of the insect gut is the careful and elegant approach taken by Beard (1960). Agents which block somatic neuromuscular function, but not visceral activity, were used to obtain paralyzed larvae of *Galleria mellonella* which contained active viscera; in effect the larvae were paralyzed in a flaccid or vegetative manner.

Larvae were injected with filtered aqueous extracts of whole *Microbracon hebetor* (=*Habrobracon juglandis*) wasps or the wasps were allowed to sting the larvae. Spider venom was also used (*Theridion tepidariorium*), and nicotine proved effective. No attempts were made to evaluate the actions of these treatments on visceral organs.

Paralyzed larvae were mounted on plastic blocks in between two electrode holders (Figure 4-25a). A blunted 30 gauge syringe needle, the recording electrode, was teased into the mouth and carefully pushed into

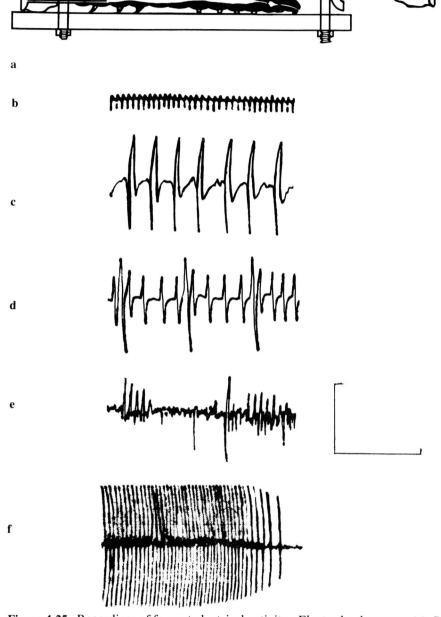

Figure 4-25. Recordings of foregut electrical activity. Electrode placement (a); P type contractions (b); V contractions (c); V contractions (larger potentials) plus X type contractions (smaller potentials) (d); a combination of P, X, and V type contractions irregular (e); rebound from a quiescent preparation to P contractions caused by the injection of 5-HT (injection syringe is shown on the right of the preparation in a) (f). Calibration: vertical 10 mV; horizontal, 1 min (b, e, f), 12 sec (d, c). After Beard (1960).

the lumen of the foregut. A second syringe needle of 23 gauge impaled the integument of a posterior segment. The wound healed quickly with little loss of hemolymph. Because of the materials used, catheters could be introduced into the electrodes to inject materials into the foregut lumen or into the abdominal hemocoel.

Quiescent preparations showed low-level electrical activity which was indistinguishable from records obtained when the recording electrode had accidentally been pushed through the foregut into the hemolymph, thus shorting the electrodes. Beard was able to initiate contractile responses with 5-HT in these quiescent preparations (Beard, 1960). Dr. Beard does not recall the concentration of 5-HT used for stimulating foregut activity (R. L. Beard, 1976, personal communication). He remembers that the effect was lasting rather than transitory and the activity introduced by 5-HT was primarily of the "P" type (Figure 4-25f).

Although foregut peristalsis was often complex, partial dissection of the larvae enabled observations of foregut contractions to be compared to recorded potentials. Four large potentials were identified in this manner with specific types of contraction.

Pulsatory proventricular constrictions immediately anterior to the midgut occurred with monophasic potentials of about 1 mV amplitude. These contractions were termed "P" and occurred up to 3 Hz but were rhythmic or irregular in character (Figure 4-25b).

Peristaltic contractions of the proventriculus which propagated anteriorly and ended at the juncture of the crop with the proventriculus were termed "V" contractions. Electrical activity accompanying V contractions were strong biphasic potentials about 10 mV or less in amplitude; however, the V contraction could be complete or abortive to various degrees, and the corresponding potentials showed various degrees of electrical activity differing mainly on amplitude (Figure 4-25c).

Simple or complex constrictions of the crop–proventricular region were termed "X" contractions. These also produced biphasic potentials and although the amplitudes of the X potentials were not consistent from one preparation to another, the X and V contractions did produce distinctive potentials which were readily discernible (Figure 4-25d).

Other contractions were observed in the *in situ* gut preparation, but those described above were the most obvious. When especially complex contractions occurred, the recorded potentials were also complex (Figure 4-25e).

Griffiths and Tauber Preparation

Griffiths and Tauber (1943) described a foregut preparation from *Periplaneta americana*. They were primarily interested in developing a saline solution based on the longevity of the foregut contractions.

The cockroach was anesthetized with ether vapors, the legs were removed, and the body was opened along the ventrum. Tracheal and tissue attachments were severed freeing the intact head and the foregut from the thorax and neck. The hindgut was severed immediately behind the gizzard. A thread was tied to the gut at the ventriculus. No difference in crop activity was noted when the knot was placed either just anterior to or just posterior to the gizzard.

Another thread was tied to the antennae or the head, so that the preparation included the whole head with foregut attached and the gizzard. Evidently this procedure including the head was inspired by Yeager's (1931) results where decapitation reduced gut activity.

The gut was mounted head up in a chamber (Figure 4-26) by looping the posterior thread around a hook in the cork in the bottom of the chamber and by extending the antenna thread out of the top of the

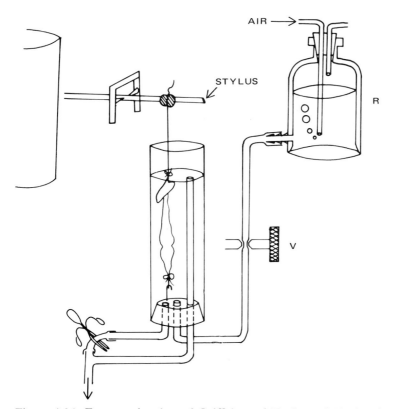

Figure 4-26. Foregut chamber of Griffiths and Tauber (1943) for the American cockroach. The head of the cockroach is intact and tied to the stylus by a thread around the antennae. Note the reservoir bottle does not maintain a constant pressure head. Air is bubbled through the saline solution (arrow). Redrawn from Griffiths and Tauber (1943).

chamber where it was attached to the stylus needle of a kymograph. Mechanical advantage was used to magnify movements by seven times. The crop chamber was a simple open tube with a paraffin-treated cork sealing the bottom. Saline was delivered through the cork from a reservoir, overflow saline was carried out of the chamber by a long tube in the chamber, and the overflow tubing also contained a second outlet to drain the crop chamber. Oxygen was bubbled through the reservoir bottle and oxygenated saline flowed at 8.3 ml/min (0.5 liter/hr) through the crop chamber.

The saline producing the best amplitude of contractions for the longest time was composed of 14.63 g/liter (250 mM) NaCl, 0.45 g/liter (6.0 mM) KC1, and 0.50 g/liter (2.0 mM) NaHCO$_3$. Crops prepared from male cockroaches produced significantly higher activity than those from females even though no special precautions or pretreatment regime was followed. Cockroaches were taken from laboratory cultures directly.

Brown Preparation

Brian Brown (1965) used a perfusion chamber for studying the action of drugs and tissue extracts on the contractions of the foregut from *Periplaneta americana*. Evidently Brown had not seen the work by Griffiths and Tauber for he made no mention of it and even used another saline, Pringle's saline with a phosphate buffer.

The cockroaches were dissected without special treatment. The foregut was exposed *in situ*, and then partially filled with saline by injection through the proventriculus into the crop. The gut was ligatured at its emergence from the head and immediately anterior to the proventriculus. The foregut was then cut away from the head and the proventriculus, and the crop was stripped of the ingluvial ganglion (Willey, 1961).

The perfusion chamber was similar to that of Griffiths and Tauber (1943) with the addition of a side arm near the top of the glass tube for runoff and a rubber stopper in the bottom. Oxygenated saline was introduced through the rubber stopper and oxygen was bubbled through the chamber constantly from a 30-gauge syringe needle which penetrated the rubber stopper. The chamber volume was 2 ml and drugs were introduced in 10 to 200 μl aliquots for assay.

Tension was about 100 mg and a stylus attached to the anterior thread on the gut wrote on smoked paper. The foregut preparation usually contracted spontaneously, but occasionally preparations were quiescent. Either preparation was extremely sensitive to 5-hydroxytryptamine at a threshold near $10^{-9}M$. Five-HT doses caused an increase in phasic contractions at low doses and tonic contractions at very high doses (Figure 4-27).

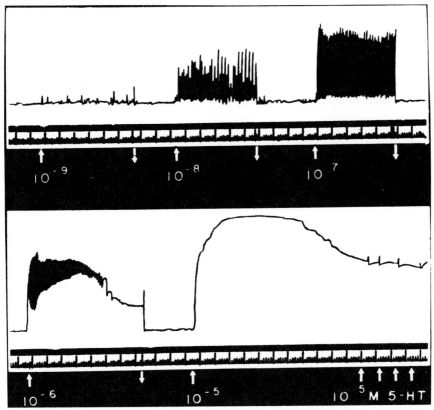

Figure 4-27. Increase in spontaneous contractile activity of the isolated foregut of *Periplaneta americana* upon the addition of increasing amounts of 5-hydroxytryptamine (5-HT). The major time notches are 1 min, subdivided into 10-sec marks. The amplitude is an arbitrary scale. These are scrapings from a smoked-drum kymograph. From Brown (1965).

Möhl Preparation

A procedure similar to that of Beard and unlike the preceding description was developed by Bernhard Möhl to study the contractile character of the cricket foregut, *Acheta domesticus* L. Möhl (1972) was interested in the neural control of foregut movements and devised a transducer to record gut movement using the Hall effect.

Adult male or female crickets were anesthetized with CO_2 and opened mid-dorsally. The head was left intact and the thorax and anterior part of the abdomen were trimmed away to expose the foregut (Figure 4-28). The foregut was then cut dorsally since contractions of the ventral part of the foregut were found to be the most consistent. Often foregut activity

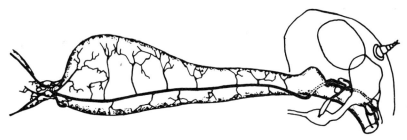

Figure 4-28. A drawing of the cricket foregut with innervation pattern. The outline of the head is shown. After Möhl (1972).

was irregular, but the ventral part of the foregut did show waves of contraction supported by the longitudinal muscles which traveled anteriorly.

These peristaltic waves were recorded with a special transducer using a Hall effect transducer (Figure 4-29). A steel wire of 100 μm diameter (Wr) was electrolytically sharpened and bent at one end. The other end of the wire held a small magnet (Mg) and the middle was fixed to two pairs of pivot bearings which allowed movement in a horizontal and vertical plane (Figure 4-29a and inset). A Hall generator (HG) was attached to a rod so as to be above the magnet and the rod was fixed to one of the pivot arms so that both the magnet and Hall generator followed vertical movement together. Horizontal movements caused the magnet to move with respect to the Hall generator producing a changing voltage on the recording apparatus. Thus the transducer followed only horizontal movements and ignored purely vertical movements. The transducer could not follow movements in line with the shaft of the steel wire.

The weight of the Hall generator was counterbalanced (Wt) to lie securely on the foregut without excess pressure and a restraining bar (B) was positioned to prevent rotation of the steel wire when removing the transducer from the foregut.

Five transducers were used at once (Figure 4-29b) and each transducer could be manipulated in three planes (J1, J2, J3) by a handle (H). The manipulators were made from brass tubes. An outer tube was slit and set over an inner tube with enough friction to hold the transducer in a steady position (Figure 4-29a).

e. Hindgut

Much research on the neural control of insect gut and on gut motility has been done on foregut, hindgut, or Malpighian tubules. Midgut has been examined very little even though midgut musculature is well developed. Therefore, midgut is not included in this treatment. Hindgut has been studied about as much as foregut, and specialized regions of the hindgut have received considerable attention, especially the rectum.

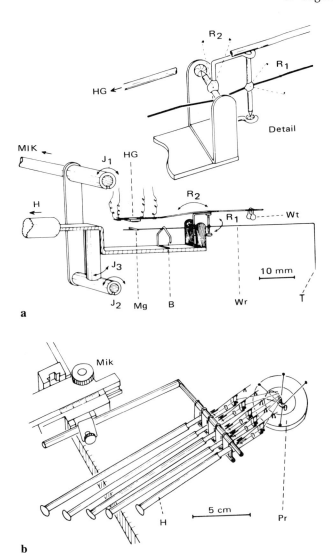

Figure 4-29. Transducers used for recording movements of the cricket foregut. (a) Detail of one transducer. The inset shows details of the movement pivot which rotates horizontally and sagittally. (b) shows five transducers ganged together to monitor five positions on the foregut. Mik, micromanipulator; HG, Hall effect generator; Wt, weight; Wr, wire probe; Pr, preparation dish; Mg, magnet; H, holders; B, wire probe restraint; J_{1-3}, manipulator movements; $R_{1,2}$, rotational movements; T, probe tip. From Möhl (1972).

Kanehisa Preparation

A preparation was developed to study movements of the hindgut of *Periplaneta americana* by Kanehisa (1965) and Brown (1965) and of

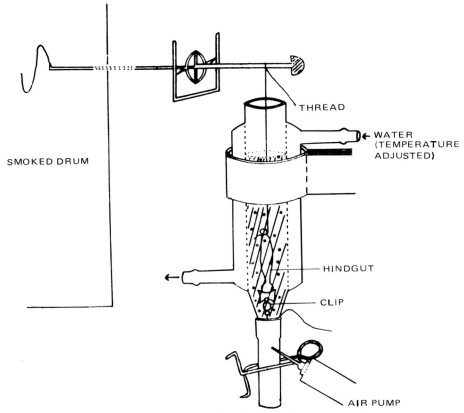

Figure 4-30. Hindgut perfusion chamber for *Periplaneta americana*. A water jacket maintains a set temperature of the saline which is oxygenated from below. The hindgut is left attached to the posterior cuticle which is clipped and holds the bottom of the gut by a thread which is pinched in the evacuation tube. The anterior end of the hindgut is tied to a kymograph stylus. From Kanehisa (1965).

Leucophaea maderae by Cook and Reinecke (1973). Kanehisa gave details of his preparation and procedures and assayed drugs (Kanehisa, 1966a, b) on the hindgut preparation from *Periplaneta americana*.

The chamber used for drug and temperature studies was a tube similar to those used for foregut perfusion, but with a modified water jacket which allowed control of the bath temperature (Figure 4-30). The top of the tube was open and allowed attachment of a gut ligature to a kymograph stylus. The bottom of the chamber provided a tube connection which was used after the gut preparation was in place and secured the bottom ligature thread.

Kanehisa used a suspension medium for his studies, not a flowing saline as had Griffiths and Tauber. However, he did use one of the latter's saline solutions: 239.3 mM NaCl, 2.7 mM KCl, 3.6 mM CaCl$_2$ · 2H$_2$O, and 2.38

mM NaHCO$_3$. This value of calcium was found necessary to maintain good peristaltic movement for long periods. The 1965 Kanehisa paper made no mention of pH; however, the pH condition was very important.

With pH adjusted to various values by 1/15 M phosphate buffer, the following types of contractile activity of the hindgut were obtained: (1) Below pH 6.0, abrupt contraction was followed by gradual extension and no further movement. (2) Between pH 6.0 and 7.5 fair movement was obtained compared to control but somewhat lower in amplitude. (3) The range of pH 7.5 to 9.0 gave good movement; used as control, the saline solution was adjusted to pH 7.8–8.2. (4) Between pH 9.0 and 10.0 fair movement was obtained with increase in amplitude and decrease in frequency accompanying the increase in pH. (5) Above pH 10.0 the hindgut contracted into prolonged tetany without further movement (Kanehisa, 1975, personal communication, Institute for Agricultural and Biological Sciences, Okayama University, Kurashiki, Japan).

A graph of the number of contractions of the hindgut per minute at various temperatures from 5 to 40°C was presented (Kanehisa, 1965; Fig. 4-31). Despite an irregular activity in the hindgut contractions and a higher frequency and amplitude of contractions at higher temperatures, a frequency of peristaltic contractions was recorded for each temperature in the range studied by neglecting small amplitude contractions. The temperatures between 22 and 32°C were considered optimal for the hindgut preparation, but the reasons were not elaborated.

Dissection was preceded by manually pushing a fecal pellet from the abdomen. The abdomen was opened and the hindgut was separated from

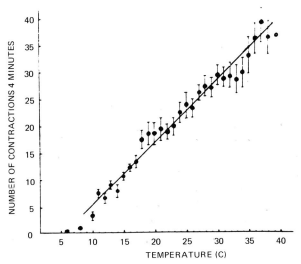

Figure 4-31. Spontaneous contractile rate of the American cockroach hindgut in a Kanehisa chamber (Figure 4-30) at various temperatures. The ordinate is number of contractions in 4 min. From Kanehisa (1965).

the surrounding tissues. The whole hindgut was ligatured and left attached to the anal cuticle. The terminal sclerite was freed with the hindgut from the abdomen and fixed by a thread and a small clip. Hindgut movement was irregular for several minutes after mounting in the suspension chamber, but later developed rhythmicity which lasted several hours.

The volume of the suspension chamber was 10 ml and aeration was found necessary otherwise activity gradually declined. Test compounds or drugs were dissolved in saline, the suspension chamber was emptied by loosening a pinch clamp on the outlet tube (Figure 4-30), the tube was resealed and the test saline introduced at the top of the chamber.

Of the compounds tested 5-hydroxytryptamine (5-HT) showed a threshold of 10^{-11} g/ml (2.47×10^{-14} M), acetylcholine increased the frequency of contractile activity at a threshold of 10^{-8} g/ml (5.5×10^{-8} M), and glutamate showed the most activity of amino acids with thresholds near 10^{-8} g/ml (16.8×10^{-8} M). The response of the hindgut to 5-HT was similar to responses of the foregut from *Periplaneta americana* (Brown, 1965), although the foregut was more sensitive.

Sakai (1973) assayed nereistoxin using the Kanehisa preparation. Unfortunately Kanehisa himself is no longer working on the hindgut nor on insect pharmacology (Kanehisa, 1975, personal communication).

Odland and Jones (1975) studied *in vivo* contractions of the hindgut of mosquitoes, *Aedes aegypti*, compared to the activity of hindguts in total or semiisolated preparations. For *in vivo* observations, mosquitoes were anesthetized with nitrogen and glued ventral-side-up onto a glass slide (Elmer's glue, Borden Co.). Scales were scraped from the abdomen and the dorsal part of the abdomen was also glued to the slide.

Semiisolated preparations were made by pulling the terminalium into a drop of saline with fine forceps. The terminal abdominal ganglia were intact and the pylorus, ileum, and colon were exposed. Hindguts were also observed after removal of the terminal abdominal ganglia.

The semiisolated hindguts contracted rhythmically in solutions of NaCl alone, but in a manner which was not comparable qualitatively with *in vivo* hindgut activity. Five grams/liter NaCl (85.5 mM) produced the strongest and most rhythmic contractions. Combining the 5 g/liter NaCl with various concentrations of KCl, it was found that 0.1 g/liter (1.3 mM) KCl and 5 g/liter NaCl produced the activity most nearly like that *in vivo*. Similarly by adding $CaCl_2$ solution to 5 g/liter NaCl and 0.1 g/liter KCl, it was found that a saline with the combination including 0.03 g/liter (0.3 mM) $CaCl_2$ produced the activity most nearly like that *in vivo* for females. Male hindguts were found to contract with the best rhythmic character in a saline with 85.5 mM NaCl, 1.3 mM KCl and 0.7 mM $CaCl_2$.

While hindgut movements were supported by many combinations of saline, the midgut ceased to contract in semiisolated preparations. This suggests that a saline developed for one organ is not necessarily ideal for others, although Odland and Jones (1975) noted "good" ovarian contractions in their best female hindgut saline. Useful comments were also made on the salines of Bradford and Ramsey (1949), Ephrusse and Beadle (1936), and Hays (1953) which were found unsuitable for supporting hindgut activity by Odland and Jones.

Brown (1965) also used a hindgut preparation from *Periplaneta americana*. He found a threshold of activity of 5-HT around 10^{-7} M on the hindgut which was prepared in a similar manner to the procedures described above for foregut. The portion of the gut posterior to and excluding the Malpighian tubules was used. He noted the hindgut underwent a gradual relaxation for 30 min after mounting in the perfusion chamber.

Kooistra (1950) found that gut preparations were poor when prepared from American cockroaches which had been deprived of water. A note was also included that the animals had to be kept at 25–27°C for satisfactory results. Presumably this meant rearing temperature. Cockroaches fed moderately and with empty or half-filled intestines produced the most satisfactory preparations. The entire gut was prepared with no details concerning the treatment of the ends. The esophagus was pinned to a wax dish, the anal end was attached to a stylus, and movements were recorded by camera as the stylus interrupted a light beam. The entire preparation lying horizontally was immersed in a temperature bath and perfused with unbuffered Levy's saline (154 mM NaCl, 9.4 mM KCl, 4.12 mM CaCl$_2$). Kooistra (1950) reported irregular rhythmic activity lasting 1-2 hr. The lack of persistent activity may be explained by the inappropriate saline. Evidently Kooistra had not seen the work of Griffiths and Tauber (1943) which was not referenced by Kooistra.

Cook Preparations

Working with his colleagues G. M. Holman, Gerald Holt, Ed Marks, and John Reinecke at the United States Department of Agriculture (USDA) Radiation and Metabolism Laboratory in Fargo, North Dakota, Benjamin Cook published a series of papers starting in 1967 dealing with tissue extracts in which the insect gut was used as an assay tool (Cook, 1967; Cook et al., 1969; Holman and Cook, 1970, 1972; Cook and Holman, 1975a, b; Cook and Reinecke, 1973; Cook et al., 1971; Reinecke et al., 1973). A hindgut preparation was also used to assay a suspected neurohormone termed hindgut-stimulating neurohormone (HSN) (Holman and Marks, 1974; Marks et al., 1973). The visceral organ preparations of Cook and his colleagues show considerable ingenuity. For example, instead of troubling to construct a device to record cockroach and grasshopper heartbeat, they simply positioned a microelectrode tip tangent to the surface of the heart tissues. The movement of the heart then bent the tip of the microelectrode to give a typical movement artifact and this was recorded as a changing resistance (Holman and Cook, 1972).

The Cook hindgut preparation was developed for *Leucophaea maderae,* the Madeira cockroach. The saline was relatively simple: 156 mM NaCl, 2.7 mM KCl, 1.8 mM CaCl$_2$, and 22 mM glucose. The pH was "adjusted" to 6.8, but no buffer was reported (Cook and Holman, 1975b).

Dissection of the gut is described in Holman and Cook (1970). Male Madeira cockroaches were decapitated, then the legs and wings were removed. The cockroaches were opened middorsally to the terminal abdominal segments. The intersegmental membranes between the eighth and tenth abdominal tergites and the seventh and eighth abdominal sternites were severed along with the minor attachments between the ventral genital sclerite and the underlying abdominal sternites 6 and 7. The male genitalia were removed. Malpighian tubules and trachea were cut from the surface of the hindgut and the hindgut was severed just anterior to the Malpighian tubules.

The last two abdominal ganglia were freed from their tracheal and nerve attachments in the ventrum. The cercal nerves containing the proctodeal nerves were carefully separated from their lateral connections. This preparation with the terminal abdominal ganglion was referred to as an innervated hindgut (Holman and Cook, 1970).

The isolated hindgut and adhering anal cuticle were transferred to a wax-filled Petri dish. Threads were tied around the anterior end just above the Malpighian tubules. A pinhole was punched through the tenth abdominal tergite and ninth sternite and a thread was fed through both sclerites and secured.

The anterior thread was tied to a hooked hypodermic needle in a gut chamber and the thread from the terminal sclerite was looped over a Palmer lever clamped in a ringstand (Figure 4-32). Thus the gut was mounted upside down. Optimal contractions were obtained with tensions of 50 mg or less. The gut chamber contained 12 ml of saline which was drained via a plug in the bottom and replenished by pouring in the top. Saline was not constantly perfused but was static with aeration.

The transducer was a Model DS 1000 linear displacement transducer with Model 201B exciter–demodulator (Daytronic Corporation). Also suitable was the Schaevitz linear viable differential transformers and amplifier–demodulators (Schaevitz Engineering, Inc.). The magnetic core element of the transducer was connected to the Palmer lever (Figure 4-32) and a counterweight was placed near the preparation to balance the weight of the transducer core.

To stimulate the nerves leading from the abdominal ganglia to the rectal muscles (via the proctodaeal nerve), a chamber was used which contained

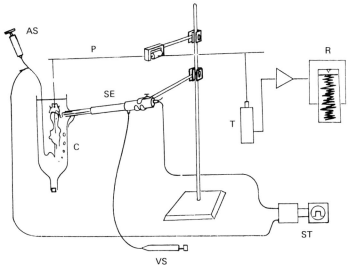

Figure 4-32. Isolated whole hindgut nerve–muscle preparation from *Leucophaea maderae* cockroach.   An aspirator syringe (AS) oxygenates the saline and provides the bottom point of attachment for the gut. The top of the gut is tied to a stylus (P) connected to a transducer and recorder (T, R). A suction electrode (SE) is sealed into the side of the perfusion chamber (C) and holds the proctodeal nerve. Stimulation is delivered between the syringe needle aspirating the saline and the suction electrode. VS, vacuum syringe for the suction electrode; ST, stimulator and stimulus isolational unit.   (B. J. Cook, 1975, personal communication.)

a 1.3 cm diameter hole near the top of the chamber tube (Figure 4-32). With the gut in place, a suction electrode was placed through the hole and sealed with dental wax. Then the abdominal ganglia were drawn into the suction electrode up to the cercal nerves. Shocks were delivered between a clip attached to the aeration syringe needle and the internal element of the suction electrode.

Hindgut preparations reportedly lasted for 8 to 12 hr. Drugs or tissue extracts were added to the suspension solution in the gut chamber and left for 5 min.   Then the chamber was flushed with fresh saline and the preparation was left 15 min between assays.

Response of the *Leucophaea maderae* hindgut preparation to L-glutamic acid at a threshold of about 10^{-5} M (10^{-6} g/ml) in the presence of 10^{-6} g/ml tetrodotoxin has suggested to Holman and Cook (1970) (and Cook and Holman, 1975a) that glutamate evoked neurotransmitter-like action on the hindgut.   However, the addition of glutamate caused an increase in frequency of postsynaptic potenials and since the glutamate response was calcium dependent, the possibility exists that glutamate is exerting its effect presynaptically (B. J. Cook, 1975, personal communication).

Figures 4-33 and 4-34 show preparations evolved for the hindgut of *Leucophaea maderae* for measuring more detailed electrical activity of the nerves and muscles. For these preparations, the gut was mounted in a horizontal bath which allowed access by recording electrodes.

For intracellular recording a floating microelectrode mounting was used (cf. Cook and Reinecke, 1973). The hindgut was prepared as above, the gut was mounted in a wax trough with a pin through the terminal sclerites instead of a thread, and a glass rod was placed in the gut lumen. A transducer probe (RCA 5734 strain gauge transducer) was placed under one of the superior rectal longitudinal muscle fibers, the fiber was lifted somewhat and impaled with the floating microelectrode.

Cook and Holman (1975a) recorded synaptic potentials extracellularly with a suction electrode from the region anterior to the rectum as shown in Figure 4-33. The synaptic potentials persisted after removal of the terminal abdominal ganglia and were abolished by perfusion with 10^{-6} g/ml tetrodotoxin. This implies a considerable amount of nervous activity associated with the hindgut quite exclusive of the central nervous system. There may be local feedback loops between stretch receptors on the gut and gut musculature, and some gut musculature appears to be electrically excitable or myogenic. The possibility also exists for motor neurons to be located in the peripheral nervous system away from the terminal abdominal ganglia and located somewhere in the cercal nerve branches or proctodeal nerves.

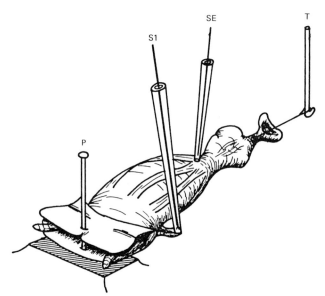

Figure 4-33. Multicellular hindgut preparation for recording junctional activity by suction electrode (SE) from *Leucophaea maderae*. T, Transducer; S1, stimulating electrode; P, restraining pin. (B. J. Cook, 1975, personal communication.)

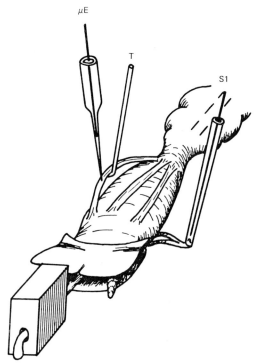

μE

T

S1

Figure 4-34. Preparation for recording intracellular potentials from superior rectal longitudinal muscles of *Leucophaea maderae* hindgut. μE, microelectrode; T, transducer; S1, stimulating electrode. Note that the glass rod inserted into the lumen of the rectum extends into the ileum. (B. J. Cook, 1975, personal communication.)

An appreciation for the complexity of hindgut nerve–muscle structure may be obtained by studying the superb description of fifth larval instar hindgut of *Manduca sexta* prepared by John Reinecke and his co-workers (Reinecke et al., 1973). Nerves innervating some of the hindgut musculature appeared to arise from nerve cells near the dorsal branch of the proctodeal nerve by the rectal valve region. Sensory neurons positioned at the anterior end of the ileum appeared to function as stretch receptors regulating muscular activity of the pylorus by sensing dilation of the ileum in *Manduca*.

While these descriptions of the innervation and control of *Manduca* larval hindgut may not be common to other insects, there is little doubt that cockroach hindgut contains neuromuscular structures of similar complexity. The interplay among intrinsic activity of the nerves and muscles of the insect gut quite apart from central nervous connection may be appreciated by the decrease in mechanical activity of the gut which is obtained on perfusion with the general nerve poison tetrodotoxin (Cook and Holman, 1975a).

Brown and Nagai Preparations

One of the most important and exciting recent findings in insect physiology was the elucidation of proctolin, a factor responsible for hindgut contraction and a suspected neurotransmitter (Brown, 1975; Starratt and Brown, 1975). The entire process from initial observations and development of a bioassay preparation to final sequencing of the peptide and synthesis started with the 1965 paper which dealt with the activity of extracts on foregut, hindgut, and heart preparations of American cockroach (Brown, 1965). Then a second paper (Brown, 1967) gave some details of neuromuscular response of the rectal longitudinal muscles of *Periplaneta americana* of the hindgut and showed that foregut and hindgut extracts provided contractions of the rectal longitudinal muscles. These extracts were termed "gut-factor" at that time.

Although Brown continued working on the gut-factor, some years passed before other work appeared. In the meantime, Toshio Nagai joined the research staff at the Research Institute, Agriculture Canada, University Sub Post Office, London, Ontario N6A 5B7, Canada. Nagai worked on the rectal longitudinal muscle preparation from *Periplaneta americana* with Brown, while Brown continued his isolation procedures on the gut-factor. From this research, several papers were published which contributed to the neurophysiology of visceral muscle (Belton and Brown, 1969; Brown and Nagai, 1969; Nagai and Brown, 1969; Nagai, 1970, 1972, 1973; Nagai and Graham, 1974). The following descriptions were taken from the series of papers published by Brown and later Nagai, and from a letter by Nagai in mid 1975.

Brown's original gut preparations were mounted in vertical gut chambers (Brown, 1965, 1967). Later preparations and most of Nagai's work used a horizontally mounted hindgut in a perfusion bath as work was focused on the rectal longitudinal muscles.

The hindgut of *Periplaneta americana* (Figure 4-35) is similar in anatomy to the hindgut of *Leucophaea maderae* (see the drawing of *Leucophaea* hindgut in Cook and Reinecke, 1973). The hindgut, posterior to the Malpighian tubules, is composed of an anterior region of circular and longitudinal muscles, and a posterior region, the rectum, consisting of a circular muscle layer which lies under six discrete bundles of longitudinal muscles. These latter are termed the rectal longitudinal muscles (Figure 4-35, Long. M.). Each bundle of the rectal longitudinal muscles consists of an underlying shorter bundle, the inferior, and an overlying bundle, the superior. The inferior straps are connected to the underlying layers of circular muscle by connective tissue over their entire length. The superior bundles are free along their entire length.

The superior rectal longitudinal muscles consist of 15 to 25 muscle fibers with an average diameter of 10 μm. They attach from the anterior of the rectum to about two-thirds the way toward the rear of the rectum

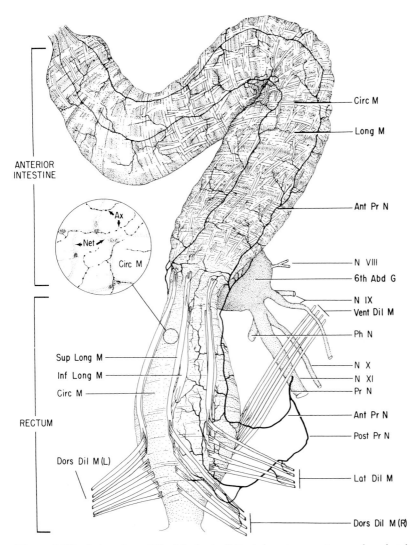

Figure 4-35. A drawing of the hindgut of *Periplaneta americana* showing innervation by branches from the proctodeal nerve from the sixth abdominal ganglion. The inset shows the surface detail from the rectum. Reproduced from Brown and Nagai (1969). Abbreviations: N, nerve; Long, longitudinal; Circ, circular; Ant, anterior; Pr, proctodeal; Abd, abdominal; Vent, ventral; Dil, dilator; M, muscle; Lat, lateral; Ax, axon; (L) and (R), left and right; Dors, dorsal.

where they give way to dilators which occupy the rear third of the rectum (Figure 4-35, Dil.). The rectal longitudinal muscles are innervated by branches arising from the paired proctodeal nerves. Each of the proctodeal nerves bifurcates into an anterior and posterior branch (Figure 4-36). The posterior branch innervates the lateral and dorsal dilator

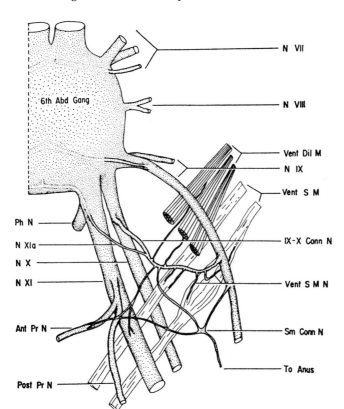

Figure 4-36. Details of the nerves leaving the sixth abdominal ganglion of *P. americana* and supplying the hindgut.   Abbreviations as in Figure 4-35. Reproduced from Brown and Nagai (1969).

muscles and the anterior circular muscles of the rectum.   The anterior proctodeal nerve innervates the ventral dilators and supplies the other areas of the hindgut including the rectal longitudinal muscles and the anterior intestine (Brown and Nagai, 1969).

Brown and Nagai (1969) found no peripheral motor neuron somata associated with the rectum in *Periplaneta*. They did find that, exclusive of longitudinal muscles dilators, the surface of the rectum was covered by what appeared to be a nerve net, but what evidently was a network of primitive muscle cells (the periproctodeal net) which receive axon endings. This network of fibers was stained with methylene blue, but its function was not known.

Two main preparations were used; either the isolated hindgut was prepared with the sixth abdominal ganglion (Figure 4-37a) or just the sixth abdominal ganglion and its innervation to a single superior rectal longitudinal muscle bundle was used (Figure 4-37b).   Later, the first

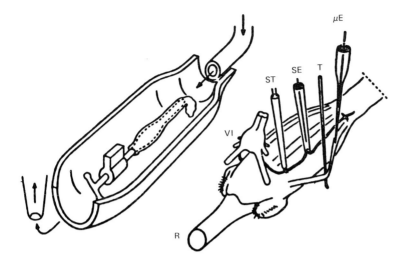

a

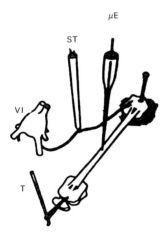

b

Figure 4-37. (a) Details of the hindgut perfusion chamber for cockroach preparations. A rod (R) is inserted into the lumen of the hindgut, the sixth abdominal ganglion (VI) is attached, and a stimulating electrode (ST) and a recording suction electrode (SE) are on the proctodeal nerve. A transducer (T) holds the rectal longitudinal muscle away from the gut and the muscle is impaled by a microelectrode (μE). (b) Detail of a single muscle preparation. Reproduced from Nagai and Brown (1969).

preparation used only the rectum (Nagai, 1973). In some cases, the sixth abdominal ganglion was removed.

The saline used was 154 mM NaCl, 2.7 mM KCl, 1.8 mM CaCl$_2$, 0.1 mM NaH$_2$PO$_4$, and 0.9 mM NaH$_2$PO$_4$ adjusted to pH 7.0. The perfusion trough held 2 ml of saline which was constantly perfused at 20 ml/min. When the rectum was used, a rotatable glass rod was inserted into the rectum (Figure 4-38). A small piece of cuticle was usually left on the posterior rectum (as shown in Figure 1 of Nagai, 1973). By holding this piece of anal cuticle with forceps, the glass rod (Figure 4-38) could be slipped into the lumen of the rectum. Nagai found the activity of the longitudinal muscle to be independent of the cuticle attached. In fact, the completely isolated preparation (Figure 4-37b) lacked cuticle.

The electrodes and transducer used with the rectal longitudinal muscle are shown in Figure 4-37a. Recording of nerve impulses from the proctodeal nerve was done with a suction electrode, and intracellular recording was accomplished with floating or hanging microelectrodes as described in Part I. An RCA 5734 or Grass Ft.03 transducer were used to record force. The muscle was stretched when necessary by a home-made device (cf. p. 170, Fig. 4-15).

The stimulating electrodes were constructed of two platinum wires insulated to the sharp curved tips by Radio TV service cement GC No. 30 Walsco 50. The tips were separated by 0.1 to 0.3 mm. Vaseline was not used to insulate the nerve and stimulating electrodes. The electrodes were always immersed in saline and a 1 msec pulse of varying voltage was used to shock nerves. It is remarkable that Nagai was able to use this arrangement without developing a stimulus artifact which obscured the recorded signal. Also, his recording arrangement (Nagai, 1973) shows the stimulating electrodes very close to the recording electrodes and this would make artifacts even more difficult to overcome. However, the beautiful recorded potentials show that the signals can be handled in this manner (see Part IV, 4, d).

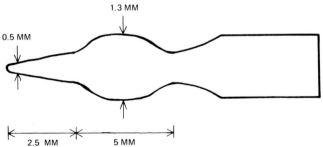

Figure 4-38. Dimensions and shape of the rod used to expand the hindgut of cockroach by insertion in the lumen of the rectum. (T. Nagai, 1975, personal communication.)

Conclusions

For whole gut preparations which are suspended vertically in a perfusion chamber, aeration appears to be absolutely necessary. The exact composition of the saline solution does not appear to be critical. Kanehisa (1965) and Brown (1965) used different salines for the same foregut preparation successfully (Table 4-1), even though both the KCl and CaCl₂ concentrations in Kanehisa's saline were more than double the values in Brown's saline.

Some authors found that handling the insects a certain way before dissection provided improved gut preparations. I suspect this may be somewhat overlooked by some workers. Thomson (1975) for example, standardized his blowflies, *Phormia regina,* in the following manner. Three days prior to experiments on *in situ* or isolated crop activity, flies were fed on 0.01 *M* sucrose for 2 days, then starved the third day. This emptied the crop (taking about 12 hr for the crop to empty). Just before experiments, flies were fed on measured volumes of 1.0 *M* sucrose containing 10 m*M* amaranth dye. Thomson found a correlation between crop volume and the contraction rate of the crop when prepared *in vitro.* He also found that the addition of 0.1 *M* glucose to saline maintained peristalsis longer than either 0.1 *M* lactose or no sugar, but that saline plus 0.1 *M* trehalose was even better in maintaining peristalsis of the crop. Thomson considered his diligent preparations necessary to obtain uniformity of response in crops.

f. Malpighian Tubules

A considerable amount of work has been done on ion transport and the excretion mechanisms of the Malpighian tubules of insects. By comparison little has been reported on the neuromuscular control of Malpighian tubule movement, and only a few articles on pharmacology have been published.

A recent study of fluid secretion by Malpighian tubules of the locust, *Schistocerca gregaria* (Maddrell and Klunsuwan, 1973), involved a simplified preparation procedure. The terminal 5 mm of the abdomen was sectioned to free the terminal gut connections, and the gut was removed by simply holding the body and pulling the head. The neck membranes ruptured and the entire gut tore free from the remaining internal tracheal and muscular connections with the body.

The gut was then submerged in mineral oil with the midgut region in contact with a bubble of saline; the severed terminus of the gut was ligated and the head tilted away to prevent regurgitated fluid from mixing with the saline bubble.

This gut procedure is reminiscent of other gut preparations. Yeager (1931) obtained his best *in vitro* foregut activity in *Periplaneta americana* by severing the neck connections and pulling the head with intact gut free from the remainder of the body. This presumably leaves the recurrent nerve and stomatogastric nervous system intact, and is similar to the operation for *Schistocerca* Malpighian tubules. Evidently an advantage accrues from leaving head structures intact in some gut preparations.

Most of the studies which have dealt with the movement of the Malpighian tubules have utilized simple observations of rhythmic contractions manually counted and timed (Palm, 1946; Koller, 1948; Meyers and Miller, 1969; Pilcher, 1971; Crowder and Shankland, 1972a, b; Flattum et al., 1973). Not all insect Malpighian tubules possess muscle fibers and many are immobile. Diptera, for example, commonly have immobile tubules while many Orthopteroid orders possess musculated tubules.

There has been a certain amount of interest in the action of tissue extracts from insects upon the movement of Malpighian tubules (Pilcher, 1971; Koller, 1948; Flattum et al., 1973). Extensive observations by Palm (1946) revealed that motile Malpighian tubules were capable of movement in simple NaCl solutions, even producing up to 12 hr of rhythmic contractions in some species. The addition of potassium or calcium salts had variable effects and provided improved longevity of contractions in the proper proportion. The addition of glucose also appeared to assist in supporting contractions for extended periods providing the salt proportions were appropriate.

Crowder and Shankland (1972a) found that the Yamasaki and Narahashi (1959) saline supported active writhing of isolated Malpighian tubules from *Periplaneta americana* for 12 hr. The saline of Palm (1946) (154.0 mM NaCl, 5.4 mM KCl, 1.8 mM CaCl$_2$, 1.2 mM NaHCO$_3$, and 1 g glucose per liter) supported active writhing for 8 hr and the saline of Miller (188.0 mM NaCl, 20.1 mM KCl, 9.0 mM CaCl$_2$, 1.0 mM unbuffered MgSO$_4$) (Meyers and Miller, 1969) did not support rhythmic contractions for longer than 5 min. Meyers and Miller (1969) reported the same result when observing activity in tubules attached to semiisolated gut preparations.

The actions of 5-HT (serotonin) on insect Malpighian tubules are instructive. Pilcher (1971) found serotonin increased the rate of rhythmic contractions of *Carausius* tubules with a threshold of near 10^{-10} M. Crowder and Shankland (1972a) found the threshold for 5-HT on isolated tubules from *Periplaneta americana* to be near 10^{-9} M and the action of 5-HT blockers suggested the 5-HT was acting at so-called D-receptors (compared to the M receptor). Neither *Carausius* nor *Periplaneta* tubules receive nerve fibers, so that the contractions are apparently an inherent property of the tubule muscle. Flattum et al. (1973) found the movements of isolated tubules of *Schistocerca gregaria* to be insensitive to the presence of 5-HT and they used this fact to distinguish between the actions of blood extracts in assays of blood-borne stress factors.

Roger Flattum and his colleagues worked out a statistical method for measuring the writhing rates of isolated Malpighian tubules; however, a more direct measure of this activity is needed. The salines needed to support extended rhythmic contractions of tubules are straightforward from Palm's work and the isolated Malpighian tubule is one of the simplest visceral muscles to prepare.

The muscle itself is extremely narrow (Crowder and Shankland, 1972b) making intracellular studies difficult. However, drug studies would be fairly simple and the tubules would be amenable for use as a bioassay for 5-HT, for example. The Malpighian tubule remains a worthwhile research problem for investigating the control of the movement of visceral organs.

g. Salivary Glands

The salivary glands of insects are either innervated or not, in a situation much like that found with Malpighian tubules. At present both innervated and noninnervated salivary glands have been examined in some detail.

Without a doubt the most famous of insect salivary glands are those of *Drosophila* which have been studied by geneticists and cytologists for many years because of the large chromosomes which are readily visible with microscopy. The larvae of chironomid midges are the source of salivary glands which have also been studied extensively because of their large size. In this case junctional connections between cells are studied since it is rather straightforward to record from several cells simultaneously (Rose, 1970).

Salivary glands are also studied for their secretion properites (for a review see Berridge and Prince, 1972). The noninnervated glands of *Calliphora* are also of interest because the lack of obvious direct nervous control suggests some form of hormonal control. In connection with the subject of hormonal control, the remarkable ability of extremely small amounts of 5-HT to increase the rate of fluid secretion in *in vitro* salivary glands of *Calliphora* has been the subject of numerous studies (Berridge and Patel, 1968; Prince and Berridge, 1973).

The effect of 5-HT on some insect salivary glands emphasizes the curious coincidental similarities between salivary glands and Malpighian tubules of insects. Salivary glands and Malpighian tubules are either innervated or not. The function of a diuretic hormone to control fluid secretion of Malpighian tubules in *Rhodnius prolixus* is perhaps the best known case of neurohormonal control of organ function in insects (Maddrell, 1971; Aston and White, 1974).

On the other hand, the work on neural control of innervated salivary glands in insects is rapidly becoming one of the best demonstrated cases of direct neural control of organ function.

Whitehead (1971, 1973) published his dissertation work on innervation of the salivary glands of American cockroach, *Periplaneta americana,* and Robertson (1974) described the innervation of the salivary gland of the moth *Manduca sexta.* The descriptions given below are from the work of Randall House and his colleagues who studied the innervated salivary glands of the cockroach *Nauphoeta cinerea* (House, 1973; Bowser-Riley and House, 1976; House, 1975; House et al., 1973; Bland and House, 1971; Bland et al., 1973; Ginsborg et al., 1974; House and Ginsborg, 1976; Ginsborg et al., 1976). *Nauphoeta* innervation is similar to innervation in *Periplaneta.* I am told *Nauphoeta* was selected for study because it produced a more acceptable odor in culture than *Periplaneta.*

The cockroaches are dissected as follows with no anesthesia.

1. Remove all legs and wings.
2. Pin dorsal side up in a wax dish or equivalent, but preferably with a black surface underneath. Place one pin in the head and, with another, stretch and pin down the posterior part of the abdomen.
3. Cut a window in the thorax with care so that the underlying gut is not damaged, and wet the tissues with saline solution.
4. Now extend the dissection by removing more of the dorsal cuticle on the thorax and abdomen. Take care not to cut the gut which may be manipulated to one side during trimming.
5. Cutting should proceed to the level of the ventral nerve cord. The salivary glands adhere to the gut and sometimes extend into the abdomen.
6. The esophagus, salivary gland ducts, and ventral nerve cord emerge from the intact head capsule; gently roll the head to one side and make an incision underneath the head from the neck to the labrum. When the latter cut is accomplished, the subesophageal ganglion is exposed.
7. The salivary gland may be isolated by cutting away the ventral nerve cord and gingerly separating the reservoir, acini glands, and salivary ducts from the gut. The saline solution was 160 mM NaCl, 10 mM KCl, 5 mM CaCl$_2$, 1 mM NaHCO$_3$, and 0.1 mM NaH$_2$PO$_4$ (House, 1973).

For some earlier studies on electrical responses of the salivary glands to nerve stimulation, the glands were mounted between platinum screens and the nerves were stimulated by field stimulation. The nerves innervating the acini glands adhere to the paired salivary ducts. Another method of stimulating these nerves has been to cut one duct anteriorly and suck both the duct and the nerve bundles into a suction electrode. Because of the poor seal developed by the electrode in this arrangement, rather large voltages have been used to evoke impulses (30 V).

A chamber was developed by Dr. House in which to mount the isolated salivary gland tissues for drug perfusion and electrophysiology

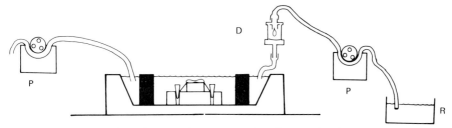

Figure 4-39. Perfusion chamber for cockroach salivary glands showing pump, P; drip chamber, D; and saline reservoir, R. (After Ginsborg et al., 1974, and C. R. House, 1973, personal communication.)

studies (Ginsborg et al., 1974). Several features of the chamber are notable and may very well be adaptable to other tissues. The basic chamber is a trough fed in this application by saline pumped in (Watson Marlow peristaltic pump type MHRE 200). The run-off is also pumped out by a second route on the same pump (Figure 4-39). The glands were draped over a central platform which was separated from each end by a nylon mesh screen. The screen had the effect of smoothing the saline flow.

By placing a drip chamber (as found on hospital intravenous tubing apparatus) in the inlet side of the perfusion tubing, interfering electrical activity from laboratory power lines was considerably reduced or eliminated. By adjusting the level of the outflow tube, the level of saline can be adjusted to the desired level over the tissues which are stretched across a platform. For mounting the salivary glands, ligatures are tied with strands of silk thread (0.007 in. size or 178 μm diameter) around the posterior ends of the paired reservoirs (Figure 4-40). These strands of thread are then pinioned at one end of the chamber by pinching the thread into one of the holes provided with the end of a wooden match stick (or round toothpick, etc.). A third ligature is tied around the anterior common point of the acini glands which is composed largely of connective tissue (Figure 4-40). The processes of pinching the final ligature can be done in such a way that the tissue is slightly stretched for most convenient access by microelectrodes.

The chamber described above has been used with double microelectrodes for measuring voltage responses to transmembrane current pulses. The mechanized pumping system has been found to be electrically "noisy" even though the saline is grounded and conducted through polyethylene or polypropylene tubing throughout. The basic problem, of course, is providing a path for high-frequency electrical activity to be carried from outside the preparation area (where the pump must be positioned) to the chamber itself where the recording electrodes pick up extraneous currents. The drip chamber eliminated most of the interference.

One clear advantage of using the House-chamber for perfusion studies

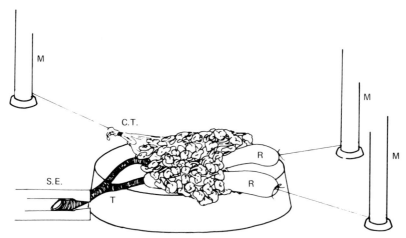

Figure 4-40. Details of the arrangement of the salivary gland tissues mounted in the House chamber (in Figure 4-39). Threads holding the salivary reservoirs, R, and the anterior connective tissue, C.T., are pinched into holes by match sticks, M, and a suction electrode, S.E., holds the anterior tracheal tubing, T, with accompanying nerves for stimulation. (C. R. House, 1973, personal communication.)

is that solutions containing drugs may be pumped into the chamber by simply removing the uptake tube from the saline reservoir (Figure 4-39), waiting for an air bubble to be pulled into the tubing, and then placing the tubing into a solution containing a drug. Thus the air gap does not allow the drug solution to mix with the saline, and the air gap can be watched to precisely time the arrival of the drug solution at the chamber. The nylon mesh screens provide a mixing action and the platform profile in the middle of the chamber ensures that the tissue comes into good contact with the drug.

Pharmacological studies to date show that perfusion by low concentrations of 5-HT or dopamine produces responses in the acinar cells of *Nauphoeta* which mimic neurally evoked hyperpolarizing responses (House et al., 1973). 5-HT also produces fluid secretion by salivary glands of *Periplaneta* at concentrations as low as $10^{-12}M$. On closer examination, however, especially when using amine localization techniques and using substances blocking synaptic effects such as pentolamine, dopamine appears to behave more like the normal neurotransmitter in *Nauphoeta* (House and Ginsborg, 1976; Ginsborg et al., 1976; House et al., 1973) and in *Manduca* (Robertson, 1975). Thus this is another example of the action of 5-HT on an insect tissue. Although remarkable in requiring very small amounts at threshold levels, 5-HT is shown not to be involved in control or regulation of the organ in any known manner in *Nauphoeta*.

h. Oviducts

Work on the physiology of oviduct neuromusculature and on the control of oviduct function has progressed at a slow pace. Since Ken Davey's review (Davey, 1964) on the control of visceral muscles in insects, a few articles have been published on oviduct preparations. Notable among these is the preparation reported by Girardie and Lafon-Cazal (1972) on contractions of the oviducts from *Locusta migratoria* in a perfusion chamber.

The oviduct chamber (Figure 4-41) was similar in all respects to the chambers used for insect gut preparations described above. Oviducts were taken from young female locusts, 12 to 15 days following the imaginal molt. The oviduct was ligatured at its two extremes and suspended in the perfusion chamber by attaching the bottom ligature to a glass aeration tube at the bottom of the chamber. The top ligature was attached to a Palmer lever.

Because of the extremely fragile nature of the oviduct tissue, a very

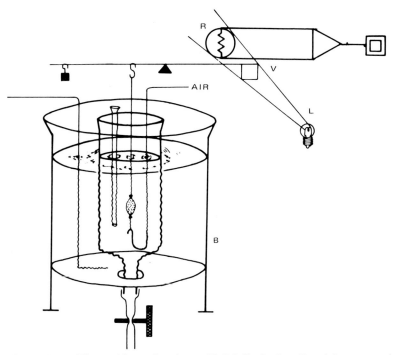

Figure 4-41. The oviduct chamber of Mirielle Lafon-Cazal is mounted inside a temperature bath, B. Air is delivered via the rod affixed to the bottom of the oviduct tissue (stippled). The top of the oviduct is tied to a stylus which moves a vane, V, in a light beam, L, to produce a transducer signal at a photoresistor, R. After Lafon-Cazal, 1974, personal communication.

sensitive transducer was employed to record contractions. The Palmer lever was balanced on a fulcrum with a vane on one side. The weight of the vane was counterbalanced (Figure 4-41, TARE) in such a manner that the Palmer lever provided minimal force on the oviduct tissue. In fact the counterbalance could be adjusted to obtain suitable excursions of the recording device during contractions of the oviducts.

The vane on the Palmer lever interrupted a light beam which was focused on a photoelectric cell. Signals from the photoelectric device were amplified and recorded.

The study using the oviduct chamber was concerned with the action of tissue extracts on the activity of the oviduct. Extracts were obtained from the pars intercerebralis, subesophageal ganglion, corpora cardiaca, and ventral nerve cord.

It has long been suspected that vitellogenesis is controlled by a neurohormonal interplay between the brain and retrocerebral complex anteriorly and the ovarian tissues in the abdomen (Nayar, 1958; Highnam, 1962; Hagadorn, 1974). The final step in this interaction is to bring about orderly contractions of the muscles of the oviducts which propel eggs in the process of egg laying (Engelmann, 1970; Thomas and Mesnier, 1973; Roth, 1973). The muscles of the oviducts of some insects are also used at times other than parturition to perform secondary functions such as shifting sperm (Davey, 1964).

The control of muscles of insect oviducts has been under sporadic examination for some time. Since tissue extracts can induce the oviduct muscles to contract rhythmically, it has been suggested that a blood-borne factor may be responsible for causing oviduct contractions at the proper time in the parturition sequence (Davey, 1964).

It is worth mentioning in this context that Highnam (1964) reported that 5-HT caused "strong and prolonged contractions of the oviducts *in vitro*" in locusts. Although it is difficult to find further reports of the action of 5-HT on insect oviducts, this one brief mention puts oviduct muscle in the same category as Malpighian tubule tissues, Malpighian tubule muscle, salivary gland tissue, some gut muscle, and heart muscle which to some degree share the common property of a sensitivity to 5-HT. The sensitivity to 5-HT is not universal since many of these tissues or muscles are not responsive to 5-HT, and which muscles or tissues are responsive to 5-HT is not readily predictable. However, the notable feature is that 5-HT can be active in extremely small concentrations regardless of the phenomenon it accelerates.

A role for 5-HT in insects has not been proven; nevertheless 5-HT has been under active investigation for some time. So far, although it appears in the nervous system (Klemm, 1976), its role as a neurohumor or neurohormone has not as yet been demonstrated.

4. Nerve Preparations

a. Introduction

The use of electronic devices to measure nerve activity in insects was applied to the sensory, muscular, and central nervous systems. Pumphrey and Rawdon-Smith (1937) and later Kenneth Roeder (1948) used these methods to great advantage in studying the neural connections between the cercal nerves and the giant axons of American cockroach. Roeder has been primarily interested in the study of insect behavior, especially its neurophysiological basis; however, he also studied the mode and site of action of DDT using early neurophysiological methods of recording peripheral and central nervous activity (Roeder and Weiant, 1946, 1948). In this study, as in many others, Roeder and his colleagues were pioneers.

From the early papers of Pumphrey and Rawdon-Smith, the cercal nerve–giant axon synapse and the ventral nerve cord itself of the American cockroach have been favorite preparations. Over the years much of the promise of this preparation originally pointed out by these authors has been fulfilled, both in analysis of the function of the nervous system and in the analysis of drug and insecticide action (for reviews see Shankland, 1976; Narahashi, 1963, 1976; Parnas and Dagan, 1971; Pichon, 1974; Callec, 1974).

The original extracellular recording methods used with the cockroach ventral nerve cord gave way to intracellular recording methods (Narahashi, 1963; Parnas et al., 1969; Spira et al., 1976) and then to oil-gap (Pichon and Callec, 1970) and to mannitol-gap (Hue et al., 1975; Callec and Sattelle, 1973; Sattelle et al., 1976).

In more recent work, the thoracic ganglia of the American cockroach are being examined at the interneuronal level with microelectrodes (Pearson and Fourtner, 1975; Pearson et al., 1976). The ganglia are also being mapped at the microscopic level (Gregory, 1974; Hearney, 1975) in somewhat of a resurgence for insect neuroanatomy. The advent of intracellular dye injection techniques has produced detailed information about the anatomy of single neurons which in turn has produced a call for more detailed information of the neuroanatomy.

The cockroach shares the honor of being one of the main experimental animals in insect nerve and muscle physiology with the locust (Hoyle, 1975). While both of the migratory locusts, *Locusta migratoria* and *Schistocerca gregaria*, have been extensively used, perhaps *Schistocerca* enjoys a slight edge in popularity for nerve–muscle work (Hoyle, 1975).

Other insects have become favorites for the demonstration of various principles in insect physiology, notably: *Rhodnius,* the blood-sucking bug; *Glossina* and other tsetse flies; *Aedes,* the malaria mosquitoes;

Calliphora, Sarcophaga, Drosophila, the fruit fly, *Phormia, Musca* among the Diptera; *Bombyx mori, Calpodes, Manduca sexta,* and cecropia among Lepidoptera; *Apis,* the honeybee; *Carausius,* the stick insect; various crickets; and the list could go on. None of these insects, except for *Drosophila,* however has been examined by neurophysiologists and neuroanatomists in anywhere near the depth that cockroaches and Orthoptera have been examined.

The main advantage of *Periplaneta* is its ease of rearing and ready availability. Availability happens to be a drawback to work on *Schistocerca gregaria,* which is restricted from being cultured in much of the United States for fear of establishing it as an agricultural pest. Under certain circumstances, males may be imported; however, a local species in California, *Schistocerca nitens* (formerly *S. vaga*), has been used as a substitute and has been cultured for a number of years by Hugh Rowell at the Zoology Department, The University of California, Berkeley, California 94720. In fact *S. vaga* overwinter on the University of California, Riverside campus, and may be obtained in February when the weather usually warms to the 80°F range for a few weeks.

The choice of experimental animal in insect neurophysiology is a matter of serious concern to some. Graham Hoyle, for instance, has generated a personal campaign behind the notion that Orthopteroid insects including grasshoppers, locusts, and crickets will yield the most realistic results in studies of neuroethology and that work on any other insect would be a virtual waste of time in the long run (Hoyle, 1975).

This will be sad news to those neurophysiologists who have invested so much effort on *Periplaneta americana.* The hard facts are, however, that *Periplaneta americana* remains the lowest common denominator for experimental animals among insects. They may be reared far less expensively than almost any other insect. One needs only a garbage can, rat food, and water. Their distribution is world-wide and they represent no threat to agriculture. The investment in time for rearing of Orthopteroids is uniformly greater, and not availabe to the solitary or casual research worker under ordinary circumstances.

Several arguments can be added to Graham Hoyle's personal preference for Orthopteroids starting with the aesthetic appeal of preferring to listen to crickets chirp rather than smelling American cockroaches in culture. For class work the grasshopper, preferably *Schistocerca,* is indispensable for demonstrating slow and fast motor units to students in introductory insect physiology. Crickets may also be used, but they are far more difficult for students to dissect because of the smaller amount of space around the metathoracic ganglion. Thus if one wants to take advantage of the worthwhile demonstration of slow and fast units, grasshoppers offer advantages.

On the other hand, the ventral nerve cord of the American cockroach is valuable on its own for classroom demonstrations of nerve impulses or synaptic transmission. Thus in these two areas of entomology teaching, arguments can be made for both cockroaches and grasshoppers.

There does not seem to be an easy answer to the problem of choosing an experimental insect species. Obviously one may choose not to work on *Schistocerca* if one is interested in sound communication among insects or in insect social behavior, or even insect toxicology or genetics. I suspect Orthopteroids would be used far more if they were made available by member governments through the United Nations or some other multinational organization such as FAO.

The original motivating force behind the stimulus to work on locusts was from agricultural economics. A need arose to somehow deal with the brutal destruction caused by migratory locust swarms mainly in North Africa and the Near East. The Anti-Locust Research Centre (Now the COPR, Centre for Overseas Pest Research) in London provided locust colonies for years to anyone interested in doing work on locusts. A similar effort was not sustained by any agency in North or South America for various reasons, and COPR has reduced its service of supplying locusts in the past several years. In North America nerve-muscle research on large grasshoppers remains possible with imported male *Schistocerca* or *Romalea microptera* or *Melanoplus* in season unless diapause can be avoided and individual colonies are maintained through the winter.

b. The Cockroach Ventral Nerve Cord

(1) Elementary Classroom Nerve Chamber

For classroom demonstrations of the conduction of the nervous impulse, refractory period, and spontaneous central nervous activity, the ventral nerve cord of the cockroach is very useful. We have used the apparatus described below in introductory insect physiology for 5 years in a 3 hr sequence which included grasshopper dissection and demonstration of slow and fast motor units.

The first 3 hr period employed a stimulator (Grass SD-9) and an oscilloscope (Tektronix 502A). Students were given the operating manuals and asked to become familiar with the controls, and then they were asked to calculate the frequency of the calibration pulse signal from the oscilloscope. The students finished by viewing the various signals from the stimulator. They compared the frequency and amplitude as given on the stimulator controls with the values recorded on the oscilloscope.

The second laboratory period was concerned with the grasshopper slow and fast motor units, and the third period used a special nerve chamber (Figure 4-42) designed to accept the American cockroach ventral nerve cord. The chamber was made from four pieces of Plexiglas at

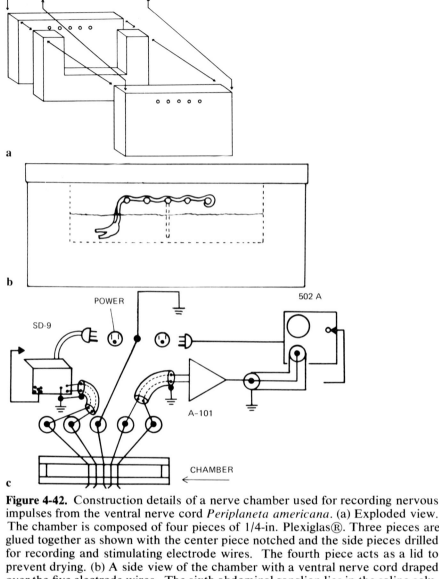

Figure 4-42. Construction details of a nerve chamber used for recording nervous impulses from the ventral nerve cord *Periplaneta americana*. (a) Exploded view. The chamber is composed of four pieces of 1/4-in. Plexiglas®. Three pieces are glued together as shown with the center piece notched and the side pieces drilled for recording and stimulating electrode wires. The fourth piece acts as a lid to prevent drying. (b) A side view of the chamber with a ventral nerve cord draped over the five electrode wires. The sixth abdominal ganglion lies in the saline solution and the center wire is connected to the saline by an extension. The solution level must be kept beneath and not touching the electrode wires. (c) A wiring diagram of instrumentation connections to the nerve chamber. Two pairs of electrodes are connected to the output of a Grass SD-9 stimulator; the middle ground is connected to a grounded screw at the power outlet; the opposite two pairs of electrodes are connected differentially to the input of an Isleworth A-101 amplifier, and then to an oscilloscope, the Tektronix 502A. The stimulator synchronous pulse (arrow) is connected to the oscilloscope external trigger (arrow). An audio amplifier should be connected in parallel to the oscilloscope input or alternatively can be connected to the oscilloscope signal output if one is available.

least 7 mm thick and with the other dimensions of about 8 × 2 cm. These pieces made the top sides and middle of a sandwich with the hollow chamber fashioned by cutting a notch of about 5 × 1.3 cm centered in the middle piece (Figure 4-42a).

After cutting the notch, the sides and center were glued together with a plastic adhesive and five small holes were drilled in both sides 2 mm from the top (Figure 4-42b). Small silver wires were mounted through the holes, glued in place, and soldered to five convenient connectors nearby. Before preparing the ventral nerve cord, saline was pipetted into the chamber to just below but not touching the wires. Adult male (usually) American cockroaches were decapitated, the legs and wings were removed, and the dorsal part of the abdomen was dissected to allow removal of the gut and exposure of the ventral nerve cord.

Only the abdominal part of the ventral nerve cord was used, therefore the nerve cord was cut through at the thorax juncture and the cercal nerves were cut. Experience showed that if the hyperneural muscle (cf. Miller and Adams, 1974) was carefully removed (an operation greatly aided by the use of watchmaker's forceps and iris scissors), the ventral nerve cord would lie across the electrodes better. About 70% of the time, a beginning student's dissection was unsuccessful so it was profitable to allow time for the preparation of the nerve cord.

Once the hyperneural muscle was removed, the nerve cord was grasped at its anterior end and lifted straight up, away from the ventral cuticle. As the cord was raised, its tracheal and segmental nerve connections were removed by cutting between the raised cord and the cuticle. It was found best to prepare the ventral nerve cord with as little attached tracheal tubing or fat body as possible. It was sometimes necessary to trim tracheal tissues if these were left attached to the nerve cord.

The nerve cord could be mounted on the recording wires (Figure 4-42b and c) in one motion, while continuing to grasp the anterior end of the cord with the forceps. When the terminal ganglion was freed of its final attachment to the ventrum, the cord was then transferred, while dangling from the forceps, into the shallow saline pool in the nerve chamber. While still holding the anterior end of the ventral nerve cord, the cord was dunked to one side of the wire electrodes, then raised straight up out of the saline until its terminal end was just in the saline and in the middle of the bath. Then the cord was moved to and draped across the electrodes so that the terminal end remained in the saline bath and the anterior end was wrapped around and underneath the last wire electrode before finally releasing the anterior end.

It was important to keep the ventral nerve cord draped across the middle of the wires and in contact with the saline at one end. The base of the wire electrodes, where they emerge from the plastic sides, could be coated with a hydrophobic material to resist wetting and therefore shorting signals. The middle electrode was extended at one side to make contact with the saline and reduce interference.

Once the nerve cord was mounted, the cover was put on the chamber to prevent desiccation and the tissue lasted at least 3 hr in constant use. Oxygenation was never needed or used. A stimulator, differential AC amplifier, and oscilloscope were connected as shown (Figure 4-42c). If the nerve recording was noisy or unacceptable, reversing the position of the recording electrodes and stimulating electrodes sometimes cleared up the problem. Thus it was important to use the same universal connectors at the chamber for all input, ground, and output connections. The instruments were always plugged into the power outlets using the ground pin on the plugs. Audio monitors were used by connecting their inputs in parallel with the oscilloscope input. Shielded cables were used everywhere but no screened cage was needed. The latter greatly facilitated access to the instruments and the preparation.

One final note on this nerve chamber concerns other cockroaches. *Periplaneta* is well suited and appears to work rather well. Larger cockroaches such as *Gromphadorhina portentosa* or *Blaberus giganteus* have longer ventral nerve cords but do not produce nervous impulses anywhere near the quality of those obtained from *Periplaneta*.

(2) Shankland Narahashi Chamber

D. L. Shankland (now chairman of the Entomology Department, Mississippi State University, Mississippi State, Mississippi 39762) has used the cockroach ventral nerve cord preparation for many years (Shankland and Kearns, 1959). Most recently he developed a preparation for use in analyzing the actions of synaptically active chemicals (Ryan and Shankland, 1971; Shankland et al., 1973; Flattum and Shankland, 1971; Shankland and Schroeder, 1973). The following description is taken from an extremely valuable handout provided by Professor Shankland to attendees at his Electrophysiology Workshop held in December 1975 at the Entomology Society of America meeting in New Orleans and is based on original designs by Narahashi (1963).

The chamber was cut from a 1/4 in. (6 mm) sheet of Plexiglas (Figure 4-43) about 3 in. (7.6 cm) × 5/8 in. (1.6 cm) by milling a rectangular hole through the center about 2.54 cm long by 6 mm wide with two rounded notches of about 3-mm diameter on one of the long sides for perfusion and aspiration tubes (Figure 4-43). A series of syringe needle tubes was cemented into holes drilled through the edge of the chamber so that they came to be horizontally across the chamber with precise dimensions as shown (Figure 4-43a). Intramedic® tubing was epoxy cemented over the second and third needle tubes, and the plastic tubing and epoxy resin was then notched to provide contact between the metal tubes and the nerve tissue as shown on the inset in Figure 4-43a. These stimulating electrodes were connected to a stimulator via a stimulus-isolation unit for shocking

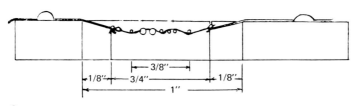

a

PLEXIGLAS NERVE CHAMBER

TO ASPIRATOR

STIMULATING
ELECTRODES
+ −

FROM SALINE RESERVOIR

REFERENCE
ELECTRODE

b

Figure 4-43. Construction details of the Shankland–Narahashi chamber for study-ing the ventral nerve cord of the cockroach, *Periplaneta americana*. (a) Arrange-ment of positioning electrodes for holding the cord and of stimulating electrodes (detail shown on inset). (b) Three-quarter view of the nerve chamber cemented to its bottom platform. A mirror may be hung beneath the platform. D. L. Shankland, 1976, personal communication.

the nerve cord. When completely milled and with electrodes and the other nerve cord support bars cemented in place, the chamber was glued with plastic cement to a larger piece of stock about 1.1 cm thick. The larger stock formed the floor of the chamber and the complete chamber was attached with machine screws to two steel arms which projected above the steel preparation plate (Figure 4-44a and b).

The entire preparation area is shown in diagram form in Figure 4-44a

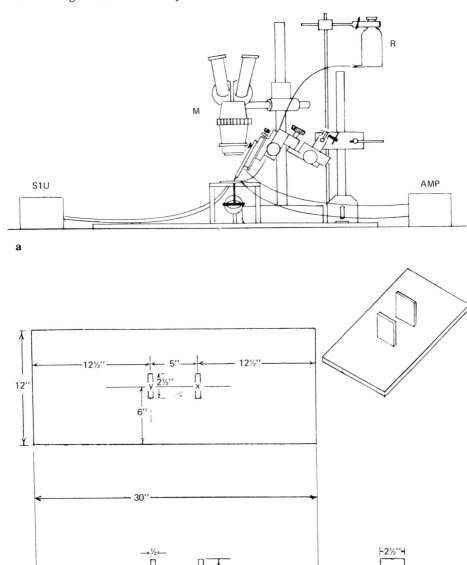

Figure 4-44. Details of the Shankland–Narahashi nerve chamber setup. (a) A suction electrode is held by a micromanipulator and connected to an amplifier (AMP) for recording nervous impulses from the ventral nerve cord. SIU, stimulus isolation unit; R, saline solution reservoir bottle; M, microscope. (b) Details of the metal plate used in the preparation with metal arms used for mounting the nerve platform. All dimensions are in inches. D. L. Shankland, 1976, personal communication.

and details of the special arms for mounting the chamber are shown in Figure 4-44b. A light source to the rear of the preparation illuminated the nerve cord from below by reflection from a microscope substage mirror mounted from the chamber platform and positioned between the steel arms of the plate holder. Also shown in Figure 4-44a are a dissecting microscope on a universal stand and a micromanipulator which could be used to hold a microelectrode or suction electrode for recording from the nerve cord or one of the ganglia. Micorelectrode penetration was facilitated by holding the nerve cord securely by threads at the ends and penetrating where the cord was held at one of the horizontal bars in the chamber.

Shankland's work was influenced by a collaboration with Professor Toshio Narahashi (moving to head the Pharmacology Department at Northwestern University's medical school in Chicago, Illinois) who also used the ventral nerve cord preparation of *Periplaneta* for numerous studies on the action of insecticides (Yamasaki and Narahashi, 1959; Wang et al., 1971; Narahashi, 1963, 1971, 1976). In fact the chamber described above was taken from one used by Narahashi and a variation on this chamber can be seen in Figure 1, p. 181 of the 1963 article by Narahashi in Volume I of the *Advances in Insect Physiology* series.

Another chamber was reported by Bettini et al. (1973) for recording activity from the cockroach ventral nerve cord in response to chemicals. Chambers are a matter of personal preference. The exact arrangement used would depend very much on what one wanted measured. Bettini's chamber provided a platform for the sixth abdominal ganglion which supported the nerve tissue against the force of microelectrode penetrations.

(3) *In Situ* Recording

Platforms are sometines placed under thoracic ganglia *in situ* to serve as a support for microelectrode penetrations. In these cases, a thin platform is manipulated underneath the ganglion of interest and microelectrodes are then driven toward the platform from above (cf. Hoyle, 1975; Hoyle and Burrows, 1973; Pearson and Fourtner, 1975).

Paul Burt (Rothamsted Experimental Station) performed minimal dissections on adult male *Periplaneta americana,* enough to expose the ventral nerve cord. He mixed fresh Vaseline petroleum jelly with fresh mineral oil about 1:1 and placed this mixture around the recording area. Then metal wires were placed through the grease mixture and in contact with the ventral nerve cord. No other insulation was found to be necessary to keep the saline from shorting the recording electrodes, and the arrangement remained stable for hours with very low noise levels. The Vaseline–oil mixture remained firm around the nerves and did not appear to affect the recordings at all (Burt and Goodchild, 1971, 1974).

We subsequently used the same Vaseline–oil mixture for grease electrodes (described in Part I) with good success. Burt (1974, personal communication) cautioned that old Vaseline not be used. Vaseline has a
tendency to turn yellow with age and Burt found that this caused unexpected nervous activity when used in extracellular nerve recordings.

(4) Sucrose-Gap Preparation

The measurement of synaptic potentials from the cercal nerve–giant axon
synapse in *Periplaneta* has been described several times (Pichon and
Callec, 1970; Callec and Sattelle, 1973; Hue et al., 1975; Sattelle et al.,
1976; Sattelle, 1976; R. J. Dowson, 1974, unpublished). The original
idea of sucrose-gap measurements was borrowed; oil was used first, and
then mannitol was found to be superior to sucrose for *Periplaneta* (Callec
and Sattelle, 1973).

The basic idea behind measurements with oil-gap or sucrose-gap is
the same as extracellular recording or intracellular recording. One part of
an excitable tissue in contact with a recording electrode must be insulated
from another part of the tissue in contact with a second electrode. If a
current path is allowed to develop between recording electrodes via the
saline solution which bathes the excitable tissues, then the potentials
recorded from the tissue will be shorted as effectively as if the electrodes
were connected directly together.

Sucrose-gap preparations with the ventral nerve cord of *Periplaneta*
provide two pools of saline in contact with the nerve cord and separated
by a pool of nonconducting solution, the sucrose or oil, which acts as an
insulator. The advantage of this arrangement has to do with the fortu-
nate anatomy of the sixth abdominal ganglion. If the sixth ganglion is fit
snugly into one pool of saline, postsynaptic potentials may be recorded
across a pool of sucrose or oil immediately anterior to the sixth abdominal ganglion. The fit of the ganglion is said to be critical for postsynaptic potentials to be recorded because of the space constant of the
giant axons. The space constant may be defined simply as the length of
axon or membrane over which a nonpropagating potential decrementally
decreases in amplitude to a value of $1/e$ of the original potential amplitude at the starting point (Aidley, 1971).

The description given below of construction of a sucrose-gap chamber
was written by M. E. Schroeder (now at Shell Development Co.,
Modesto, California 95352) in 1975.

(a) *Tools required.* Three miniature end mills, 1/32, 1/16, and 7/64 in.
diameter and one high-speed drill No. 57 (0.044 in.) are required. All
milling operations were done on a Unimat miniature lathe with the head in
the vertical position and operated at 2500 rpm.

(b) *Reference.* Callec and Sattelle (1973).

(c) *Procedures.*

Step 1. Cut a Plexiglas block to about $12 \times 50 \times 25$ mm. Make sure that the top and bottom of the block are smooth and parallel. Surface if necessary.

Step 2. Using the 1/32 in. end mill, cut a groove along the long axis of the block about 3/4 in. long and 3/32 in. deep (Figure 4-45).

Step 3. Again using the 1/32 in. end mill, cut a groove perpendicular to the first and about 9/32 in. from the edge of the block and make it about 1/8 in. deep.   Make sure the left-to-right displacement of the Unimat is tight during this operation. This compartment will henceforth be referred to as *Compartment 2.*

Step 4. Using the 1/16 in. end mill, cut a groove parallel to and to the right of Compartment 2.   The distance separating these two compartments is critical and should be between 1/64 and 1/32 in. Make this compartment 1/8 in. deep and extend it out to about 1/4 in. from the edge of the block.   This compartment will be *Compartment 3.*

Step 5. Again using the 1/16 in. end mill, drill two holes at each end of Compartment 2 so that the holes form expanded ends of the compartment as shown in Figure 4-45.   Drill the hole to the same depth as the compartment (1/8 in).

Step 6. Using the 7/64 in. end mill, make a groove parallel and to the right of Compartment 3.   The separation between the two compartments

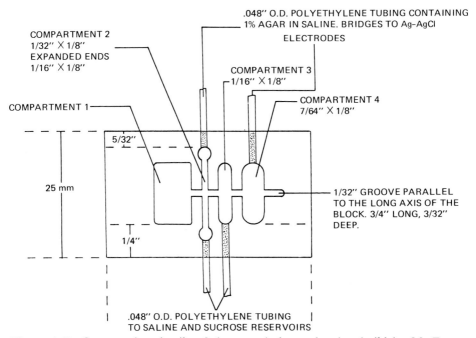

Figure 4-45. Construction details of the mannitol-gap chamber build by M. E. Schroeder (1975, unpublished).

should be 1/64–1/32 in. It should be 1/8 in. deep and extend out to 1/4 in. from the edge of the block. This is *Compartment 4*.

Step 7. Using the 7/64 in. end mill, cut a groove parallel and to the *left* of Compartment 2. Make it sufficiently large to accommodate the cercal nerves and suction electrodes. It should be 1/8 in. deep and separated from Compartment 2 by 1/64–1/32 in. This is *Compartment 1*.

Step 8. Reposition the chamber in the Unimat vise so that the compartments face you with Compartment 1 on the left (Figure 4-45). Using the No. 57 (0.044 in.) high-speed drill bit, drill two holes from the side of the chamber through to Compartments 2 and 4. These holes will receive the saline–agar bridges to the silver–silver chloride electrodes.

Step 9. Rotate the chamber 180° so that Compartment 1 is on the left. Using the same drill bit, make two more holes connecting Compartments 2 and 3 to the exterior. These holes receive the tubing from the elevated saline and sucrose reservoirs and accommodate standard 0.048 in. O.D. polyethylene tubing.

(d) *Notes on procedure*. Before using the chamber, give all of the compartments a light coat of Vaseline. This helps to break the surface tension of the saline and sucrose (or mannitol) and helps maintain a constant fluid level in each compartment. Make sure the Vaseline seals remain secure. This is especially important with respect to the sucrose (or mannitol) compartment. Before inserting the nerve into the chamber, construct the Vaseline seals. This is accomplished with a Vaseline-filled syringe equipped with a 30 gauge needle. After the seals have been constructed, start the saline and sucrose perfusion to make sure there are no leaks. A small leak can be tolerated at this time since with time the seals have a tendency to "seat" themselves more securely and small leaks are sealed.

Separate the connectives between A-5 and A-6 to facilitate penetration of the nonelectrolyte. When the nerve is ready to be placed in the chamber, remove about half of each Vaseline seal. Place the nerve cord in position with the cercal nerves in Compartment 1, the sixth abdominal ganglion in Compartment 2, and the anterior part of the nerve cord in Compartment 4. Push the nerve cord down slightly into each remaining bit of Vaseline and then apply more Vaseline to the top of the nerve to complete each seal. Start the perfusion slowly. Try to keep the nerve below the surface of the saline and sucrose. This helps alleviate some of the stress on the seals. Fill the other two compartments with saline. Allow about 40 min for the sucrose (or mannitol) to penetrate the connectives after which time a stable excitatory postsynaptic potential (EPSP) should be evoked by stimulation (Figure 4-46a). Without stimulation, spontaneous potentials are recorded whose origin is not entirely clear, but may represent action potentials in some of the smaller units in the nerve cord (Figure 4-46b).

Upon stimulation of the cercal nerves, as the voltage is slowly increased, an EPSP is recorded (Figure 4-46a). As the stimulus intensity

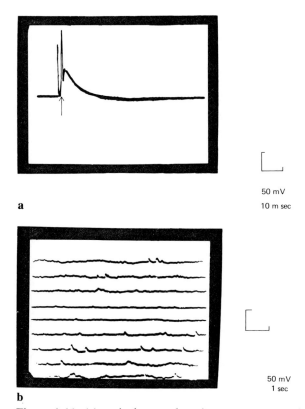

50 mV

a   10 m sec

50 mV
b   1 sec

Figure 4-46. (a) typical control excitatory postsynaptic potential as recorded with sucrose-gap technique. The first spike is a stimulation artifact and the second spike (arrow) is thought to be a spike in the cercal nerve. (b) Spontaneous potentials recorded from sucrose-gap preparation. Calibration: 50 mV, 1 sec (b), 10 msec (a). M. E. Schroeder, 1975, personal communication, unpublished.

is increased more axons in the cercal nerves are recruited and these produce a larger complex EPSP. At still higher stimulus intensity, the EPSP summate to produce an action potential which appears as a spike on the top of the EPSP. This is distinct from the spike recorded from cercal nerve action potentials (Figure 4-46a, arrow). As the stimulating voltage is increased higher than threshold for spikes, compound action potentials are recorded on the EPSP (Callec and Sattelle, 1973).

The cercal nerves may be stimulated by a suction electrode or by a pair of wire electrodes (as described by Callec and Sattelle, 1973 and Hue et al., 1975). The perfused saline and sucrose (or mannitol) were pulled off by suction. The saline solution used by Callec and Sattelle was 708.6 mM NaCl, 3.1 mM KCl, 5.4 mM CaCl$_2$, and 2.0 mM NaHCO$_3$ adjusted to pH 7.0. The mannitol solution was 87 g/liter.

Others have used sucrose-gap chambers for recording synaptic events.

R. J. Dowson (when at the Shell Research Labs, Sittingbourne, Kent, England) simply placed the sixth abdominal ganglion in a saline/drug perfusion chamber, with the anterior ventral nerve cord draped through ordinary high-vacuum silicone grease and into a second saline chamber. Dowson stimulated the cercal nerves by a suction electrode.

Mrs. Geescke Friedricksen (Sektion Biologie–Tierphysiologie, Friedrich-Schiller-Universität, 69 Jena, East Germany) simply placed the excised ventral nerve cord across three solutions on a microscope slide. The sixth abdominal ganglion rested in a saline pool and the fifth abdominal ganglion was placed in a second saline pool with the intervening connectives in a sucrose pool. The solutions were formed and defined by ridges of Vaseline. The nerve cord was pinned anteriorly and posteriorly into very small patches of wax on the slide. The slide was mounted under a dissecting microscope with a black stage to cut one connective down to the giant axons in the sucrose chamber. Air was bubbled through the sixth abdominal ganglion chamber and the saline chambers were covered with a film of oil.

c. Extracellular Recordings from Peripheral Nerves

To a person facing a need to record action potentials from insect axons for whatever reason, there are few resources readily available with practical descriptions of what to do. Recording potentials from muscles with wire electrodes placed through the cuticle is rather straightforward (and was described in Part III). It is possible to record nervous impulses from whole limbs without dissecting the cuticle away, but many nerve recordings are done in preparations which are soaked in saline solution and represent certain challenges in recordings.

Fortunately a few publications concerning insects have taken the time to describe extracellular nerve recordings in detail among which are Singh et al. (1972) and Pearson et al. (1970). I have found the latter article particularly helpful, and this will be reviewed here briefly.

Extracellular electrodes consisted of 75 μm diameter (about 3 mil or 0.0029 in.) silver wires. One silver wire was shaped in a hook and manipulated under the axon bundle of interest. The bundle was pulled above the hemolymph or saline and the exposed portion was painted with Vaseline petroleum jelly to prevent drying. A second silver wire electrode was placed near the first in the hemolymph. When amplified (by connecting differentially to the input pins of an Isleworth amplifier for example), this gave a triphasic record for nervous impulses. Small nerves can be damaged by this procedure and cautions were given to move the nerves as little as possible, preferring to suck the hemolymph level down by syringe rather than raising the nerve bundle away from the tissue.

By cutting the axon bundle described above just distal to the recording electrodes, the potentials became diphasic but then shortly developed into monophasic nerve pulses. Pearson et al. (1970) also described a method for measuring the resistance in the nerve bundle between the recording electrodes. They merely connected a variable resistance between the electrodes at the amplifier input. Starting at about 1 MΩ, when the resistance was decreased to the point where the recorded nerve impulse amplitudes were about half of the unshorted values, this value of resistance approximated the source resistance. They found less than 200 kΩ for monophasic recording and less than 100 kΩ for triphasic recording.

Nervous impulses recorded extracellularly are typically in the range of a few hundred microvolts to a few millivolts. To predict the success of recording with various combinations of instruments, the specifications listed for the amplifiers can be compared to the amplitude of the nerve potentials recorded. The noise level of the Isleworth A-101 AC amplifier, for example, is given as 10 μV peak-to-peak with the input shorted and no filters used. If the amplifier is connected to 100 kΩ and the signals are passed between 200 Hz and 5 kHz, the noise is 13 μV peak-to-peak. Thus the Isleworth can easily distinguish a nerve impulse of 100 μV amplitude from a tissue presenting 100 kΩ impedance to the recording electrodes. The Grass P511 AC preamplifier also lists a similar noise level: 13 μV peak-to-peak with a band width of 30 kHz.

The use of oil or Vaseline in conjunction with recording nervous activity from axon bundles in insects is fairly well established now. Paul Burt (1974, personal communication) used a 1:1 Vaseline to paraffin oil mixture, Singh et al. (1972) used a 2:3 mixture, and Pearson et al. (1970) used Vaseline alone.

In their study of innervation and control of fibrillar flight muscles in the bumblebee, *Bombus* sp., Ikeda and Boettiger (1965) removed the sternites from the ventral part of the thorax, destroyed the thoracic ganglion by heat, and mounted nerve branches supplying the flight muscles on 10 μm diameter nichrome wires. The preparation was coated with heavy mineral oil to prevent drying and to insulate the wire electrodes which were used for stimulation. Electrical activity was recorded intracellularly from the flight muscles using microelectrodes.

As described in Part III, nervous activity may be recorded from axon bundles in intact limbs by a judicious selection of recording site. Either a hole is made in the cuticle and a wire is inserted near the nerve of interest, or a sharpened metal electrode may be pushed through the cuticle and used for the recording without prior preparation.

The latter method was used by Vincent Salgado (1976, unpublished) to record ascending nervous impulses from the house fly leg (Figure 4-47). The metathoracic leg from *Musca domestica* females was cut near the

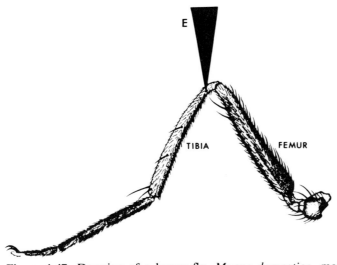

Figure 4-47. Drawing of a house fly, *Musca domestica,* metathoracic leg with sharpened tungsten electrode (E) in position for recording sensory nervous impulses from the tarsal region.

thorax and a Minuten Nadeln stainless-steel insect pin was pushed into the femur from the proximal end. This tended to block the stub of the femur and retarded drying. A sharpened tungsten pin was manipulated into the tibia just below the femur joint.

A similar recording method was developed using an isolated leg from *Periplaneta americana* (Chapman and Pankhurst, 1967; Chapman and Nichols, 1969). Pairs of stainless-steel insect pins spaced 2 to 4 mm apart served as recording electrodes and were mounted in a square array. Electrodes were placed on a piece of Plexiglas and the leg was mounted on the pins. In practice, once the leg was mounted, the connection to the recording amplifiers could be switched from one pair of pin electrodes to any other combination desired.

Nervous impulses were initiated by mechanical stimulation of campaniform sensillae associated with the tactile spines on the tibia (Chapman, 1965). Conduction velocities were measured by comparing nervous impulses recorded from two pairs of electrodes on a dual-trace oscilloscope display and by dividing the time lapse of travel from one electrode to the other into the distance between them. Impulses could be initiated by moving one of the large tactile spines on the tibia, and the oscilloscope trace could be set to trigger a sweep internally from the channel recording the first nerve impulse. Recordings presented in the Chapman and Pankhurst article show a very acceptable diphasic nerve impulse recording with peak-to-peak amplitude of 200 μV or less.

Either of the two methods of recording nervous impulses described above could be adapted to any insect with limitations in using small appendages being determined by the ability of the recording electrode to penetrate the appendage properly. In fact this same principle is used in recording from single units on insect antennae (Morita and Yamashita, 1959).

d. Stimulus Artifact Suppression

One of the most annoying realities in electrophysiological recording is that of the stimulation artifact. Other than fiddling with the geometry of the stimulating electrodes or taking pains to ensure that both stimulating and recording electrodes are insulated from the conducting medium, there are a few tricks one can resort to in seeking to reduce the artifact potential from recorded electrical activity (Roby and Lettich, 1975; Coombs, 1965).

When a ground wire was placed near the stimulating electrodes, Iles (1972) found that the artifact could be reduced by connecting a large capacitance between the bath and the ground connection. Toshio Nagai (see Part IV, 3, e) placed stimulating electrodes near recording electrodes. He was able to reduce the stimulation artifact by placing a second pair of stimulating electrodes near the first in the saline solution (Figure 4-48). By delivering a shock to the first pair of stimulating electrodes which initiated a nervous impulse and applying the same electrical shock pulse, but of opposite polarity to the second pair of electrodes, he was able to reduce the artifact substantially. In effect the artifactual electrical signals reaching the recording electrodes canceled each other. This manipulation reportedly depended somewhat on the geometry and spacing between the two pairs of stimulating electrodes. Each new preparation required some experimentation to determine the most favorable position for the electrodes (Toshio Nagai, 1976, personal communication).

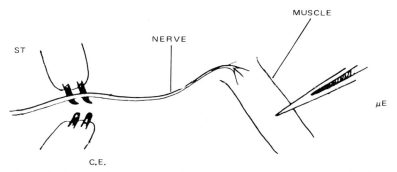

Figure 4-48. Stimulating electrodes (ST) and compensating electrodes (CE) positioned to reduce stimulus artifact from microelectrode intracellular recordings (μE). T. Nagai, 1976, personal communication.

e. Artificial Respiration for CNS Preparations

Since the advent of dye-filling techniques which have allowed both the recording of electrical activity and then the marking of neurons in the central nervous systems (CNS) for identification and anatomy, neurophysiology studies have advanced an enormous amount in a few short years. In a few insects, notably the grasshopper and the cockroach, the position and responses of almost all of the motor neurons have been determined (Hoyle, 1975). The next great challenge is in recording at the interneuron level and a few studies have already approached this (Pearson and Fourtner, 1975; Burrows, 1975; Burrows and Siegler, 1976). Techniques used in central nervous recording and dye marking are to be covered in Volume 2 of this series and therefore discussion of them is deferred; however, a note is included here concerning aeration of insects dissected and prepared for various CNS studies.

Among the most delicate neurophysiology work done on the central nervous system of insects is that accomplished by Kazuo Ikeda (Neurosciences Division, City of Hope, Duarte, California 91090) (Ikeda, 1977). Dr. Ikeda found in his work on the bumblebee that an artificial supply of air had to be supplied to the tracheal system to maintain viability (Ikeda and Boettiger, 1965).

In *Drosophila* artificial respiration consisted of attaching an air tube to the spiracles (Ikeda and Kaplan, 1970). More recently this method has been replaced by providing an air flow past the semiintact thorax which appears sufficient for maintaining aeration of the CNS tissues (Ikeda, 1974, personal communication).

We have copied Dr. Ikeda's respiration method for studies on house fly CNS. The house fly preparation consists of a Plexiglas block which is drilled and tapped in one end to provide for fitting of an adapter and an air tube. The drilled hole meets a cavity milled in the top surface (Figure 4-49a). The cavity is covered by a small plastic mounting card, and the card has a hole cut in the middle where the house fly is positioned.

House flies were anesthetized with CO_2 and placed in the hole in the mounting card, and wax was pushed up to the thorax with a warm wire. Once the thorax was waxed to the mounting card, the house fly was held for the insertion of wire electrodes into the dorsal thorax to record flight muscle potentials (cf. Part III, 4 and Figure 4-49b).

---➤

Figure 4-49. Mounting the house fly for central nervous system studies. (a) A block with air tube connected to provide artificial respiration. The electrode lead wires are seen next to the tubing. (b) Wire electrodes shown implanted into thorax dorsolongitudinal muscles. (c) House fly card turned over and house fly ventrum removed to expose the thoracic ganglion (dissection necessitates removal of the front two pairs of legs).

a

b

c

When the wires were waxed into the thorax and satisfactory recordings were obtained, the mounting card was turned over so the house fly was ventral-side-up. The ventrum of the thorax was removed including both front and middle pairs of legs and the furca-sternite to expose the compound thoracic ganglion (Figure 4-49c). A small pool of saline was placed on the exposed tissues to retard drying and microelectrodes were placed into the ventral surface of the ganglion to record activity from central neurons while recording muscle potentials from the dorsolongitudinal flight muscles. Using the same arrangement, various poisons or neurotransmitter chemicals were iontophoresed or perfused into or onto the thoracic ganglion and the responses of the flight motor neurons were observed (Miller, 1976).

The preparation described may be adapted to other Diptera or other insects depending on the information sought. Where laboratory air supplies were not available, air was supplied by pumps designed to aerate fish tanks.

5. Nerve–Muscle Preparations

a. Introduction

Much of our knowledge of insect nerve and muscle physiology comes from larger insects, especially grasshoppers, locusts, and cockroaches. A considerable amount of attention has been directed to two muscles in the locust, the retractor unguis and the extensor tibia or the jumping muscle. Possibly more is known of cockroach and locust neuromusculature than that of other insects except for the important studies on insect fibrillar flight muscles.

There have been numerous reviews on the subject of neuromuscular physiology (Faeder, 1968; McDonald, 1975; Usherwood, 1969, 1974; Hoyle, 1974; Smyth et al., 1973) and an entire book was devoted to the subject, published in 1975 (*Insect Muscle,* edited by Peter Usherwood, Academic Press, London).

Three preparations will be described below. The fast and slow units of the locust extensor tibiae will be described from Graham Hoyle's pioneering work. Then the inhibitory units of the coxae of the cockroach from work by Pearson, Iles, and Fourtner will be described and finally a fairly detailed account of the locust retractor unguis will be included because of the enormous attention given it by Professor Peter Usherwood and his students. As an added portion of this section, reference will be made to neuromuscular preparations which have been used for research on the mode and site of action of venoms.

b. Fast and Slow Motor Units

Demonstration of fast motor units can be accomplished in almost any insect. One merely mounts the decapitated subject dorsal-side-down, secures the appendages except for one rear leg, and opens the thoracic box over the metathoracic ganglion. Stimulating electrodes are then manipulated or fitted around the crural nerve (the large nerve trunk leaving the ganglion and entering the femur) leading to the unsecured leg. Shocks are delivered via paired stimulating electrodes, or a single wire may be mounted on the crural nerve and shocks delivered between it and an insect pin in the neck stump or elsewhere in the body. The latter is allowed as long as electrical activity is not being recorded from the same animal and therefore the stimulation artifact is not a factor.

To demonstrate the slow and fast contraction of the jumping leg in grasshoppers or locusts, obtain a fairly large specimen if possible, decapitate, and place the body in a shallow groove cut in a wax-filled Petri dish. Plasticine may be used instead of wax, but I prefer wax to hold the insect. Place an insect pin through the neck stump to secure the anterior end and either remove the front two pairs of legs or secure them along with the hind femurs with straps of Plasticine formed by rolling a bit between the thumb and index finger. Some find it advantageous to avoid restricting movement of the abdomen. Without disturbance, the abdomen will maintain ventilation movements which normally maintain the tracheal system filled with air.

A ventral view of the locust or grasshopper shows the appearance of the ventral cuticle and the position of the metathoracic ganglion (Figure 4-50a, dashed lines). The ganglion rests forward of the position of the rear coxae so the basisterna of the metathorax must be removed to expose most of the metathoracic ganglion. A forward incision may be made just posterior to the prominent furcal suture running transversely between the coxae of the middle legs using a new safety razor blade. If a square is cut on the basisterna and the patch of cuticle is removed, the metathoracic ganglion will be located amid a brightly colored fat body in the center of the cavity produced.

Once the pair of anterior connectives leaving the anterior end of the ganglion and the much smaller diameter connectives leaving the posterior edge are located, the large crural nerve should be found. The crural nerve in *Schistocerca* is prominent, even surprising, in that its diameter is on the same order as the large anterior connective bundles. It leaves the ganglion at an angle and travels almost straight for the cavity of the coxae of the leg it innervates.

Sometimes a drop or two of vital stain increases the contrast of the nerve tissues to assist in finding the proper nerve branches. We have used

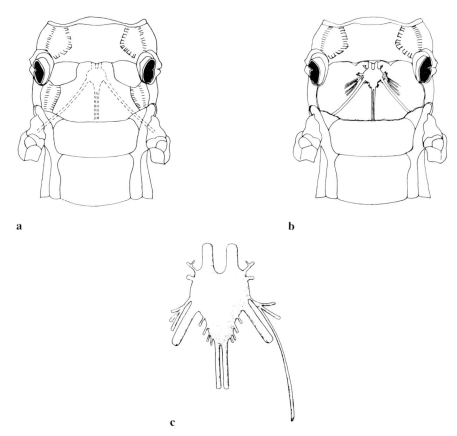

a

b

c

Figure 4-50. A drawing of the ventral aspect of the grasshopper metathorax and the metathoracic ganglion. (a) Appearance of the ventral sclerites before dissection with the dashed indicating the ganglion position. (b) After removal of the ventral cuticle to expose the ganglion *in situ*. (a) and (b), after Albrecht (1953) with permission. (c) Detail of the metathoracic ganglion with nerve branches.

0.05% Janus Green B in any general saline with good success. The stain does not appear to affect responses to stimulation adversely; however, when any saline is added to the dissected cavity, it tends to seep away immediately. It is necessary to periodically moisten the exposed nervous tissues to prevent drying, although, in some preparations, the metathoracic ganglion is destroyed to prevent spontaneous activity from causing leg movements beyond those being studied.

A wire is brought near or placed around the crural nerve and a shock may be delivered between this and the pin in the neck stump. While gradually increasing the stimulating voltage, at one point the appropriate tibia will twitch rather violently one-for-one following the delivery of each shock.

If for some reason some other muscle is observed to twitch first as the voltage is increased, this simply means the stimulating electrode is not near enough to the crural nerve for specificity. One can reposition the stimulating electrode until at some threshold value only the extensor tibia muscle is caused to contract so that the tibia twitches. The crural nerve contains numerous axons. The largest axon in diameter belongs to the so-called fast axon. This is named by virtue of the contractile response which is elicited from stimulation of the axon and does not have anything to do with propagation velocity; in fact, propagation velocity in this largest diameter axon (the fast axon) is coincidentally not significantly greater than propagation velocity in the axon of the next smaller cross-sectional diameter (one of the slow axons) (Hoyle, 1955a).

The fusion frequency of the extensor tibia muscle may be measured. By steadily increasing the frequency of stimulation at the threshold voltage, the unimpeded tibia may be seen to twitch and relax as two discernible events. At a certain point at higher frequencies, the cycle will become a fast vibration as the response follows the stimulation frequency. Then at still higher stimulation frequencies, the vibration amplitude decreases until individual movement is no longer discernible. This is the fusion frequency or the muscle is said to be in tetany and is ordinarily maximally contracted. When the stimulation frequency is increased, sometimes the threshold for stimulation of some axon other than the fast unit is reached and the tibia, tarsus, or even the femur may be moved in an unexpected direction. Again this is largely a matter of the position of the stimulating electrode.

To demonstrate the slow units supplying the extensor tibia muscle, one may take advantage of a quirk in the innervation which is peculiar to grasshoppers and their cousins. The axons which innervate the same extensor tibia muscle and cause slow responses to stimulation, leave the ganglion via nerve 3b (Figure 4-50b; locust nerve numbering as in Figure 4-52b). To demonstrate this response, cut the crural nerve and move the stimulating electrode to the next anterior bundle of axons. If the middle branch of N3 (N3b) can be located affix the electrode here, otherwise stimulation may be delivered to the entire N3 nerve trunk near its origin on the lateral aspect of the ganglion (Figure 4-50b).

Set the stimulator to deliver shocks at between 20 and 40 Hz. Then while repetitively stimulating, slowly increase the stimulating voltage. At threshold the tibia should effect a slow smooth extension. Once this occurs, the stimulation frequency may be varied and the tibia may be brought to any position between rest and near maximum extension by simply varying the frequencies from low to high.

The slow unit innervating the extensor tibia muscle is the more difficult to locate. However, its illustrative value is enormous. Students can be observed in the classroom studying the slow movement of the extensor tibia at length once the demonstration of slow movement is attained. Somehow the simple lesson behind "slow axon" seems to penetrate more in this preparation than in most. It may have something to do with the impression made as a tibia is made to respond robot-like while a dial on a machine is turned. In any event the demonstration is far more satisfying than the capricious responses encountered with taste experiments or experiments with the heartbeat for example.

c. The Inhibitory Synapse

Inhibitory transmission at the insect nerve–muscle synapse was described in the much referenced work by Usherwood and Grundfest (1965). This 1965 article and Usherwood's 1968 article on inhibitory synaptic transmission have placed the subject on a firm footing and have brought out many similarities between insect and other arthropod inhibitory nerve–muscle synapses (Hoyle, 1974).

A further similarity between crustacean and insect was pointed out by Pearson and Bergman (1969) who reported common inhibitory neurons in cockroach and locust based on electrophysiological evidence. These preliminary results were followed up by cobalt back-fills which modified the earlier findings on cockroach in which two additional inhibitory neurons were found (Figure 4-51) (Pearson and Fourtner, 1973).

A considerable amount of work has been done on the muscles of the American cockroach coxae and their role in normal walking. The original work of Carbonell (1947) described the muscles and Pipa and Cook (1959) located and labeled the nerves. Besides work on innervation of the extensor tibia (Atwood et al., 1969), Kier Pearson and his colleagues have investigated the role of the coxal units in walking in addition to sorting out the details of innervation patterns of the motor units (Pearson and Iles, 1971; Pearson and Bergman, 1969; Pearson and Fourtner, 1973, 1975; Pearson, 1972; Pearson et al., 1976; Wong and Pearson, 1976). As a result of these important advances the neuromusculature of the cock-

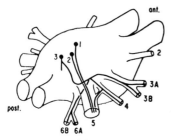

Figure 4-51. More recent details of the three common inhibitory neurons in the metathoracic ganglion of cockroach, *P. americana*. From Pearson and Fourtner (1973).

roach coxae is known in some detail. Some of this is given in Guthrie and Tindall (1968).

A similar body of knowledge has been developed for the grasshopper or locust. Hoyle (1966) presented a detailed description of the anterior coxal adductor muscle of *Schistocerca gregaria, S. vaga, Romalea micraptera,* and *Melanoplus differentialis.* The coxal adductor muscle of the locust is innervated by both excitatory and inhibitory nerves. Thus an inhibitory nerve–muscle preparation can be developed from either cockroach or grasshopper. Inhibitory postsynaptic potentials neurally evoked in the locust coxal adductor muscle are sometimes hyperpolarizing when recorded *in situ,* but they can be depolarizing when recorded *in vitro* (Usherwood, 1974) and therefore the locust preparation is somewhat unpredictable. Inhibitory postsynaptic potentials recorded from coxal muscles of the cockroach have always been reported as hyperpolarizing regardless of dissection procedures.

Since the inhibitory neuron in the locust metathoracic ganglion is a common inhibitory neuron (Figure 4-52b), once the coxal adductor is prepared in dissection, the anterior connective may be shocked to produce a nerve impulse specifically in the inhibitory axon. This is one of the methods used by Usherwood in his early studies of inhibitory synaptic transmission (Usherwood and Grundfest, 1965).

In practice, shocking any connective of an intact ganglion also produces impulses in excitatory axons as well. Some patience may be needed to produce the proper orthodromic impulse from such antidromic stimulation. Another alternative is to shock another of the ipsilateral peripheral axons hoping to evoke an impulse in the inhibitory axon supplying the coxal adductor.

The latter approach is the choice of Pearson and Bergman (1969) (Figure 4-53). Muscles 182C and 182D are the coxal levators and are located

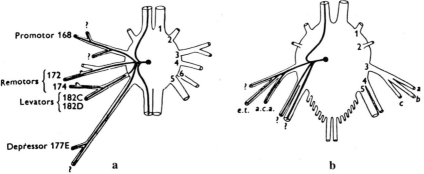

Figure 4-52. Cell body and axon positions of the common inhibitor neurons of the metathoracic ganglia of cockroach (a) and locust (b). From Pearson and Bergman (1969).

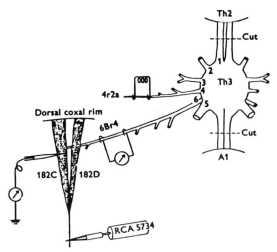

Figure 4-53. Preparation for producing inhibitory nervous impulses in the coxal muscles of *Periplaneta americana* by reflex stimulation of the common inhibitory pathway. From Pearson and Bergman (1969).

just beneath the dorsal cuticle of the coxae. They receive three inhibitory axons, the common inhibitor already described in Figure 4-52a (Pearson and Iles, 1971; Pearson and Fourtner, 1973) and two additional neurons termed the localized common inhibitory (Pearson and Fourtner, 1973) (Figure 4-51) whose branches are found only descending the crural nerve (N5). Pearson and Fourtner (1973) were unable to back-fill common inhibitory axon branches in the connectives leaving the metathoracic ganglion. They found that only ipsilateral branches of the neurons filled with cobalt. However, stimulation of the connective at high frequencies caused a 1:1 production of impulses in the inhibitory axons (Pearson and Bergman, 1969); thus there appears to be uncertainty about the presence of axonal branches in the anterior or posterior connectives from the main common inhibitor neuron. This explains the slight discrepancy between Figures 4-51a and 4-52a.

My own personal experience with the cockroach coxal nerve–muscle preparation is that antidromically eliciting impulses in the common inhibitor is somewhat tricky. We had poor luck with shocking the connective and resorted to recording synaptic potentials which were evoked from spontaneously occurring motor neuron activity. Impulses in the inhibitory axons could be evoked by tickling the tarsi and unguis of the leg being examined and inhibitory postsynaptic potentials could then be monitored for a period of time (Olsen et al., 1976).

Another minor annoyance has to do with orientation. Except for Pearson and Bergman's helpful drawing (in Figure 1 of their 1969 article), I have always found it difficult to remember exactly where the cockroach coxal muscles are and which are their proper numbers. Also when an isolated leg is presented, sometimes the authors fail to make it absolutely clear which part of the leg is up or down, in or out. For this reason the above-mentioned figure is included here for reference (Figure 4-54) with the added note that metathoracic muscles 182 lie on the *dorsal* part of the coxae and 177 lie on the *ventral* part of the metathoracic coxae. The corresponding muscles in the *mesothoracic* coxae are numbered 140 and 135 (Carbonell, 1947; Pearson and Iles, 1971; Guthrie and Tindall, 1968).

The cockroach inhibitory nerve–muscle preparation may be minimally dissected and the metathoracic coxal levators 182C and 182D may be exposed by removing a patch of cuticle from the *dorsal* part of the coxa. If the cockroach is mounted dorsum down, this means the leg being examined must be folded forward in an unusual position (Figure 4-54).

Another alternative in the dorsum-down preparation is to affix the appropriate leg with its normal ventral surface facing up (Figure 4-55) and then use the hatchet method where overlying musculature is removed including muscles 177 until the coxal remotors are exposed from inside out. One hundred eighty-two C and D usually lie flat against the cuticle. In this preparation there is the choice of recording from 182A, B, C, or D.

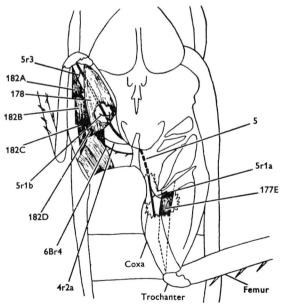

Figure 4-54. Ventral view of the American cockroach showing the position of coxal muscles and innervation in the metathoracic segment. From Pearson and Iles (1971).

Figure 4-55. Photograph of a simple cockroach coxal muscle preparation. The head, abdomen, and front two pairs of legs have been removed.

In Figure 4-55 the head, abdomen, and pro- and mesothoracic legs have been removed. The cockroach is mounted with its ventral surface up in a shallow wax bottom trough. Clips hold the pronotal shield and the right metathoracic leg. A small mound of wax was molded around the coxa–femur joint of the left metathoracic leg with a warm soldering iron. The ventral part of the coxa was removed along with much of the underlying musculature. The microelectrode is shown impaled in muscle 182 which lies in the bottom of a cavity formed from the dissected coxa with a wax mound on the sides.

In the preparation shown, drugs were perfused on the coxal muscles by flooding the cavity with drugs in saline solution. To prevent drugs from acting on the thoracic ganglia, a Vaseline bridge was formed around the inner aspect of the coxa between muscle 182 and the thoracic ganglia. A perfusion tube delivered saline to the thoracic tissues and a reference electrode was placed near the coxa.

Fourtner (1976) reported a useful observation on the appearance of various muscle fibers of the cockroach coxal muscles. He found that the fibers of metathoracic coxal muscles 178 and 179 were all innervated by the fast axon and had short sarcomeres in the range 3.2 to 3.9 μm. The fibers of muscles 177d' and 177e' had longer sarcomeres of 4.0 to 4.7 μm and were innervated by both fast and slow motor neurons. Finally, the fibers of muscles 177d and 177e were innervated by three inhibitory axons and a slow axon and had long sarcomeres between 5.7 and 8.7 μm. Since sarcomeres are visible in the freshly dissected muscle tissue, this correlation may assist in finding particular motor units from the bulk of the muscles present.

The innervation of the cockroach metathoracic extensor tibia muscle has been described in detail by Atwood et al. (1969), and a diagram was given by Cochrane et al. (1972) (Figure 4-56) which shows where fast, slow and inhibitory postsynaptic potentials can reasonably be expected to be located. A similar pattern of innervation holds for the locust extensor tibia muscle with various regions being innervated in different patterns (Figure 4-57) (Usherwood, 1967; Cochrane et al., 1972; Burrows and Hoyle, 1973; Hoyle 1955a, b).

Most of the extensor is innervated by a fast axon. At the proximal end near the coxa, a small group of fibers receives slow and inhibitory axons and another adjacent group of fibers receives slow and fast axons. A few fibers receive fast, slow, and inhibitory innervation. Walther and Rathmayer (1974) report that some of the ventral fibers on the distal end of the

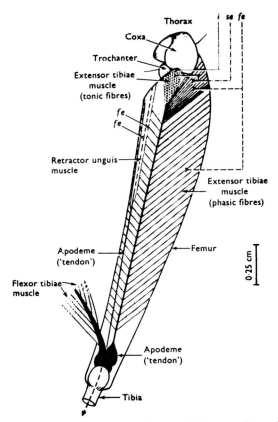

Figure 4-56. The locust femur of the metathoracic "jumping" leg showing types and position of muscles with the type of innervation. The flexor tibia muscle is not included. From Cochrane et al. (1972).

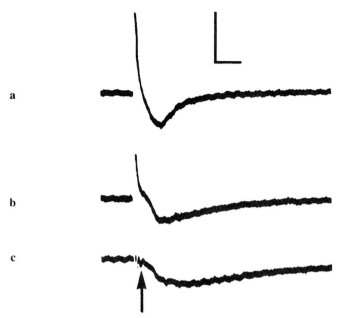

Figure 4-57. Intracellular potentials from the sternopedal muscle of *Philosamia cynthia* mesothoracic segment. (a) Neurally evoked complex postsynaptic potential from shocking the anterior connective of the thoracic ganglion. Intermediate (b) and final (c) potential thought to be an inhibitory postsynaptic potential produced over 20 min upon poisoning the muscle with extract of braconid venom apparatus. From Piek and Mantel (1970a).

extensor tibia of the hind leg of *Locusta migratoria* receive inhibitory innervation.

Thus the extensor tibia muscle is also a candidate for the inhibitory nerve–muscle preparation. I prefer the cockroach coxal muscles because the chances of impaling the proper fiber are greater, and with triple inhibitory innervation there are more inhibitory synapses to deal with. The locust flexor tibia of the metathoracic leg is less well studied (Brookes and Werman, 1973).

d. Venom Assays Using Nerve–Muscle Preparations

The venom or extracts of venom glands or even whole bodies have been assayed on several nerve–muscle preparations including some using insects. The process of poisoning by braconid wasps, *Microbracon* (= *Habrobracon*) *hebetor*, was originally investigated by Raimon Beard (1952) who recently retired from the Connecticut Agricultural Experiment Station. Tom Piek has studied the mode of action of this venom extensively with his colleagues at the University of Amsterdam (Piek and Mantel, 1970b; Piek, 1966; Piek and Engels, 1969; Visser et al., 1976; Piek et al., 1974; Drenth, 1969, 1974a, b).

For studying the action of venom extracts on insect excitatory and inhibitory nerve–muscle synapses, Piek and Mantel (1970a) chose the Saturniid moth *Philosamia cynthia*. The selection of experimental animal was somewhat biased in that the prey of the braconid wasp are normally lepidopteran larvae, although the venom does act on locust (Walther and Rathmayer, 1974).

Adult female *P. cynthia* were decapitated and dissected to expose the mesothoracic ganglion and sternopedal muscle, named II St$_1$ based on the anatomical study of Nüesch (1953). This muscle was chosen because of its apparent similarity to the mesothoracic coxal adductor muscles of Orthoptera. All mesothoracic ganglion nervous connections were severed except for those to the sternopedal muscle. Shocks were applied to the anterior connective of the mesothoracic ganglion on the assumption that a branch of a common inhibitor axon allowed antidromic impulse transmission into the ganglion then orthodromic transmission descending along an inhibitory axon to the sternopedal muscle in a manner analogous to the suspected case for locust common inhibitors (Pearson and Bergman, 1969).

This preparation in *P. cynthia* allowed evoked postsynaptic potentials (PSPs) to be recorded from the sternopedal muscle. These responses could be somewhat complex due to a simultaneous recording of excitatory (EPSP) and inhibitory (IPSP) PSPs (Figure 4-57a) depending on the responses evoked by shocking the anterior connective of the ganglion.

Perfusion of the sternopedal muscle with extract of the braconid venom apparatus caused the intracellular responses to change over 20 min (Figure 4-57b and c) such that the suspected EPSP component was markedly reduced leaving a postsynaptic potential which closely resembled simple IPSP responses (Usherwood and Grundfest, 1965). This corresponded with the known selective action of the extract at blocking excitatory neuromuscular transmission in insects at a presynaptic site without affecting inhibitory nerve–muscle synaptic transmission. This was generally confirmed by Walther and Rathmayer (1974) in locusts using the retractor unguis or extensor tibia nerve–muscle preparations from the hind leg. At present it has not been reported exactly which presynaptic mechanism in the excitatory nerve terminal has been affected: neurotransmitter synthesis, transport, or release.

The action of extracts from black widow spider venom apparati (*Latrodectus sp.*) has been examined on the metathoracic flexor tibia muscle of *P. americana* cockroach (Ornberg et al., 1976; Griffiths and Smyth, 1973) and on the metathoracic extensor tibia and retractor unguis muscles of locust, *Schistocerca gregaria* (Cull-Candy et al., 1973).

In each case the extracts of the venom apparati caused a great increase in the frequency of miniature postsynaptic potential events (mPSPs) of the excitatory synapses studied indicating a release of neurotransmitter. The release was followed by an equally drastic reduction in frequency accompanied by a block in neurotransmission.

Application of respiratory poisons or uncouplers of oxidative phosphorylatin (and other biologically active molecules such as quinones) also caused a great increase in mPSP frequency then a decrease and block in synaptic transmission at cockroach retractor unguis muscles (Miller and Rees, 1973). While there seems to be little doubt that the respiratory inhibitors acted at mitochondria in the nerve terminal, it is less obvious where the spider extract or quinones are acting. Cull-Candy et al. (1973) reported clumping or reduction in numbers of synaptic vesicles in locust motor nerve terminals and their micrograph suggests a disruption in mitochondrial cristae. A very similar description came from ultrastructural examination of cockroach retractor unguis motor–nerve terminals following blockage by respiratory poisons or quinones (Miller and Rees, 1973).

The preparations used for analysis of the action of black widow spider venom are possibly the simplest type to approach experimentally. Because of the multiterminal nature of innervation of insect muscle fibers, the chances of recording mPSPs from any skeletal muscle are excellent. The only exception is insect fibrillar flight muscle where nerve terminals are much less accessible.

The test of whether a nerve–muscle junction has become blocked is equally straightforward. One merely needs to bring two wires near the muscle in a bath, turn off the microelectrode amplifier, and shock once with a few volts. In normal muscle the shock will depolarize nerve terminals and cause a twitch assuming some fast axons are present. After (or during) poisoning, the twitch response may be checked as required. The threshold for eliciting a response directly from the muscle fibers by such an external shock is thought to be considerably higher than that necessary to depolarize the nerves present; thus this is rarely a drawback and the gross shock is a valid test of synaptic transmission.

Because the nerve terminal is the most susceptible and labile element in the nerve–muscle preparation, care must be taken to avoid a general debilitation caused by a restriction in the supply of oxygen to the tissues. As with cockroach heart, one may watch that trachea remain air-filled, or bubble oxygen or air in the bath (generally not possible near a microelectrode impalement), or ensure a flow of oxygenated saline.

The following procedure in working with retractor unguis muscle in a perfusion bath was taken from the work of Douglas Rees (personal communication, 1973) and is instructive. The retractor unguis muscle from *Blaberus giganteus* was bath mounted and the tracheal supply was removed (by cutting away all large tracheal branches). Air was bubbled in the saline bath, the nerve was shocked with a suction electrode, and twitch contractions were observed.

When the bubbling was discontinued, the rheobase for neurally evoked responses slowly increased, and then electrical and mechanical responses were lost. At this time, the mPSP events were still recorded and the membrane potential of the muscle remained unchanged. When air bubbling was restarted, all electrical and mechanical events returned and rheobase returned to its original value.

The procedure outlined above for *Blaberus* leading to blockage of nerve–muscle transmission was repeated with a retractor unguis muscle from *Periplaneta*. The *Periplaneta* synaptic transmission did not fail upon eliminating air bubbling in the perfusion bath. Thus differences between muscles will be encountered.

For references to other venom studies using insect nerve–muscle preparations see Parnas et al. (1970) who examined scorpion venom on locust extensor tibia of the metathorax. Parnas also examined the action of venom on membrane impedance by passing current and recording voltage changes simultaneously. Rathmayer (1962), Piek (1966), and Piek et al. (1971) have found a presynaptic action on locust nerve–muscle assay by digger wasp venom, *Philanthus triangulatum*. However, they also found the intriguing result that the same venom does not act on *P. triangulatum* itself.

Findlay Russell (1977, personal communication, University of Southern California Medical School, Los Angeles, California 90007) has repeatedly warned that most animal venoms are a complicated combination of biochemicals and that they rarely act at one specific site or by one mechanism alone. He cautions that some articles report the action of black widow spider venom, when actually an extract of the venom gland was used, and that in such a procedure the actual venom may constitute only 10% of the crude extract. For most researchers, however, these cautions are pedestrian since one is normally interested in the specific action of a particular fraction regardless of the method of isolation. Venom components in the past have proven of enormous value in blocking highly specific physiological functions at the membrane level, thereby providing a rare and invaluable glimpse at the underlying processes involved.

e. The Retractor Unguis Preparation

The retractor unguis preparation was developed as a model muscle for use in studying excitatory neuromuscular transmission in insects by Peter Usherwood. This muscle was found suitable because it received innervation by two excitatory motor axons of the fast variety without inhibitory innervation and without innervation by slow axons. There are actually two parts of the muscle, a section of "red" fibers and a separate and distinct bundle of "white" fibers. The red are thought to be phasic and the white tonic (Usherwood, 1967; Cochrane et al., 1972).

Thus the retractor unguis "white" fibers are innervated by a single fast axon which branches to supply multiterminal innervation along all of the fibers comprising the muscle bundle. This simplicity of innervation marks the retractor unguis as a rather unusual muscle; however, it is typical of most insect skeletal muscles in that the majority of the synaptic contacts are located on the inner aspects of the fibers. The synapse membranes are therefore relatively inaccessible unless microelectrodes are positioned into the clefts between the fibers. There are synapses on the surface of some fibers which are accessible but hard to find.

(1) Dissection Instructions

The dissection instructions given below were from extensive notes by Douglas Rees. While the drawings are of a generalized right grasshopper leg, they are applicable to cockroach and other insects. It is known that the cockroach retractor unguis (*Blaberus giganteus* and *Periplaneta americana*) is similar in appearance and innervation to the grasshopper; however, the innervation of the retractor unguis from the stick insect *Carausius morosus* is different, involving more than fast axons (P.N.R. Usherwood, personal communication, 1974).

Step 1. Remove the right hind leg by cutting through the coxa. Figure 4-58a shows the outer aspect of the right leg and Figure 4-58b shows the inner aspect of the same leg with the relative position of the extensor tibia, flexor tibia, and retractor unguis muscles in the femur. Arrows in Figure 4-58b show the approximate position of cross-sectional views used in Figures 4-59b and 4-61b.
Step 2. Make an incision along and on both sides of the femur about the middle of the herringbone pattern. In Figure 4-59a and b, XX' shows where to cut on the inner aspect of the leg and YY' shows about where to cut on the outer aspect of the leg. The herringbone pattern in grasshoppers is the site of attachment of the extensor tibia muscle. This incision may be made with small scissors or preferably with a sharp scalpel or razor blade, but must be cut superficially so as not to damage underlying tissues.
Step 3. Lift the distal end of the dorsal part of the right femur with a forefinger (Figure 4-60) and cut away the extensor tibia muscle (ET in Figure 4-59b) from underlying tissue starting with the apodeme (Figure 4-58b). Care should be taken not to cut too closely to the ventral tissues in the femur.
Step 4. Place the leg in a perfusion chamber so that the exposed tissues face downward, the ventral cuticle of the femur is up, and the tibia is also pointing up out of the perfusion chamber. Refer to Table 4-2 or to any

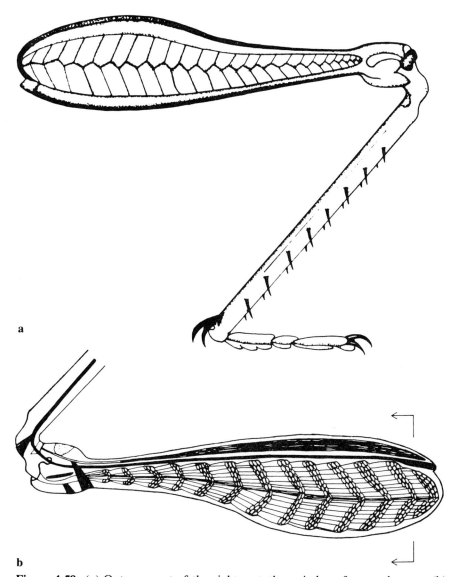

Figure 4-58. (a) Outer aspect of the right metathoracic leg of a grasshopper. (b) inner aspect of the right leg with the relative position of extensor tibia, flexor tibia, and retractor unguis muscles in the femur. Arrows show the position of cross-sectional views in Figures 4-59b and 4-61b.

recent study for the saline. The femoral–tibia joint and the coxa are secured to the bottom of a perfusion chamber with Tackiwax before wetting the tissue with a few drops of saline solution (Figure 4-61a). Care must be exercised not to heat the leg excessively to avoid damage to the

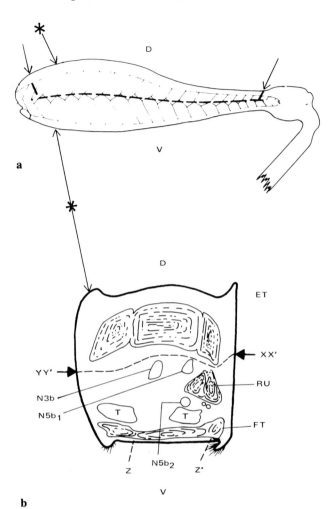

Figure 4-59. (a) Inner aspect of the right metathoracic femur of a grasshopper. Arrows mark the position of initial incisions which extend along dashed lines. (b) Cross-section of the femur near the proximal end, as indicated on Figure 4-58b, and indicated by asterisks on Figure 4-59a. Incisions of the femur along the inner (XX') and outer (YY') aspect of the cuticle, and later along the ventral, V, aspect (ZZ'). ET, extensor tibia muscle; RU, retractor unguis muscle; FT, flexor tibia muscle; T, large tracheal tubes; N3b, N5b$_1$, and N5b$_2$ are various major nerve trunks. The two small unlabeled circles ventral to RU are the two axons supplying RU.

tissues inside. It helps to have both transmitted and reflected illumination hereafter.

Step 5. The tibia is secured at a 90° angle to the femur if the retractor unguis is to be at rest length, or the tibia may be cut and the stub folded

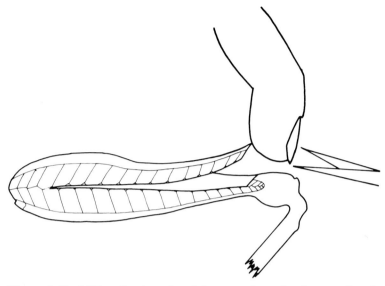

Figure 4-60. Lifting the dorsal cuticle away from the femur, after the initial circumcision of the cuticle, and trimming adhering tissue away from the inner aspect of the cuticle being removed.

right over and waxed onto the ventral part of the femur if the retractor unguis is to be stretched.

Step 6. All of the ventral aspect of the cuticle of the femur is now cut to expose the underlying tissues. The final appearance of the retractor unguis is greatly exaggerated in Figure 4-61a with the flexor muscle not shown.

Step 7. Remove the ventral cuticle of the femur. The flexor tibia may be pulled away by grasping its apodeme (taking care to miss nerve N5) and pulling up out of the femur cavity in a manner similar to that used to remove the flexor. Begin rapid and continuous perfusion of the freshly exposed tissues.

Step 8. The outer wall of the femur may now be removed (area shown dotted on Figure 61a). What remains is shown in cross section in Figure 4-61b and includes the two distinct portions of the retractor unguis muscle one of which appears pale, W, and the other pink, R, two prominent tracheal tubes [only one is shown (TR)], and the crural nerve trunk N5b$_2$, which supplies the retractor unguis muscle; two small fast axons may be present which innervate the unguis and the inner aspect of the cuticle of the femur is still present (shown as a solid line).

Step 9. The retractor unguis may be cleaned up further with the following cautionary notes: the insertion of the retractor unguis muscle lies at the trochanter end of the inner wall of the femur (the last piece of femur cuticle remaining). Nerve bundles N3b and the remaining branch of N5b which continues through the femur may be removed. The tracheal supply

Table 4-2. Experimental Salines for Retractor Unguis Preparation[a]

Species	Ionic composition (mM)								mOsm	pH	References
	Na	K	Ca	Cl	HPO$_4$	H$_2$PO$_4$	BES:NaOH	Sucrose			
Locusta migratoria or *Schistocerca gregaria*	156	10	2	154	6	4	—	—	332	6.85	Usherwood and Machili (1968)
S. vaga	160	10	2	174	—	—	1.0:0.10	12.5	360	6.90	Rees (1973, unpublished)
Periplaneta americana	190	15	4	213	—	—	—	—	422	—	Rees (1974)
P. americana	190	15	4	213	—	—	1.0:0.12	26.4	450	7.15	Rees (1973, unpublished)
Blaberus giganteus	190	10	4	208	—	—	—	—	412	—	Rees (1974)
B. giganteus	190	10	4	208	—	—	1.0:0.12	—	413	7.15	Rees (1973, unpublished)

[a]Notes. Use of a BES Good's buffer alleviates the calcium phosphate precipitate which occurs in phosphate buffers used in many insect physiological salines. Aeration of salines used for *Blaberus giganteus* did not significantly alter physiological responses of the nerve–muscle preparation when tracheal supply was left intact and attached to the retractor unguis muscle. Addition of 10 to 20 units of streptomycin–penicillin maintained nerve–muscle preparations from *P. americana* and *B. giganteus* in good physiological condition for periods of 24–30 hr with no substrate added.

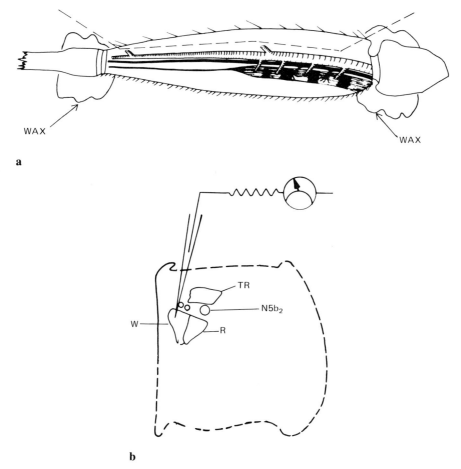

Figure 4-61. (a) View of the ventral aspect of the right femur as it is positioned dorsal-side-down in the perfusion chamber. The dashed line shows additional trimming of the remaining cuticle. (b) Cross-sectional view of the same muscle. The dorsal part of the femur is mounted down in the chamber and microelectrodes record from the unguis muscle from the ventral direction with respect to the muscle. TR, tracheal tube; $N5b_2$, crural nerve trunk; W and R, white and red muscle fibers. Dashed lines show cuticle removed during dissection.

must be left intact to the retractor unguis as well as its innervating branch of N5b. If the retractor unguis muscle is touched with dissecting instruments or pieces of cuticle, or if it is pulled unduly during dissections, the tissue may be injured. Damage can appear as ripples or bulges on the fibers during equilibration.

Step 10. The final step is to equilibrate the muscle fibers. The perfusion rate may be reduced to 0.5 ml/min and a period of 1 to 2 hr is suggested to allow equilibration between the muscle interstitial fluid and the experimental saline.

(a)

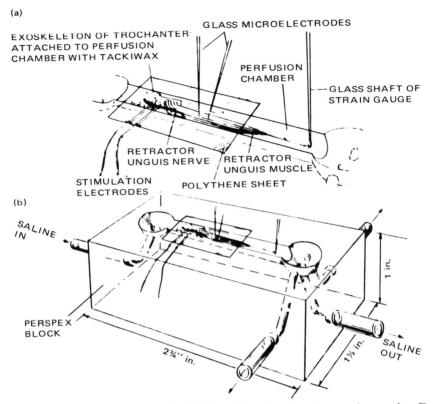

Figure 4-62. Perfusion chamber designed for the retractor unguis muscle. From Usherwood and Machili (1968).

The perfusion chamber used by Usherwood and Machili (1968) is instructive (Figure 4-62). All of the procedures outlined above were employed with this preparation except that the apodeme of the retractor unguis muscle was affixed to a transducer and the remaining cuticular tissue around the femur–tibia joint was removed. As this nerve–muscle chamber evolved after 1968, the three exit tubes were replaced by a single one and the polythene sheet was not used. Its original intention was to reduce movement caused by buoyancy of the muscle in the perfusing saline and cause the saline to wet the entire muscle by total immersion. Also, later versions were tapered from a deeper groove at the saline input to a shallower groove at the output end, or the reverse of the condition shown. In this way a constant saline level is automatically maintained over the muscle and the entire groove is accessible to electrodes. Finally, suction electrodes were found more useful than the arrangement shown for stimulation.

The following notes from Joseph P. Hodgkiss (Glasgow University,

Scotland, personal communication, 1973, and Usherwood, 1972) on isolating synapses for focal recording are helpful. To find the sites of origin of miniature postsynaptic potentials (mPSPs), the retractor unguis muscle is prepared and the nerve is shocked with about 1.5–2 V, at 0.05 msec duration and 0.5 Hz. After the muscle twitches normally in response to the shocks, Mg saline is added at a concentration which just eliminates movement from shocks.

Normal locust saline for these studies was 140 mM NaCl, 10 mM KCl, 4 mM NaH$_2$PO$_4$, 6 mM Na$_2$HPO$_4$, and 2 mM CaCl$_2$. To reduce or eliminate movement 40 to 50 mM Mg was added to the saline with enough Na removed to rebalance the osmolarity. Dowson and Usherwood (1972) added 15 mM Mg and reduced the calcium from 2 to 0.5 mM to decrease the amplitude of neurally evoked EPSP responses below threshold for activation of nonsynaptic muscle membrane in the locust metathoracic retractor unguis muscle (Usherwood, 1972).

After eliminating twitches, while still shocking, impale a fiber with one microelectrode to record a diminished EPSP (Figure 4-63d). Move a second extracellular recording microelectrode along a cleft alongside the same muscle fiber (shown in diagrammatic form in Figure 4-63b). Neurally evoked potentials of 500 μV amplitude may be recorded extracellularly near synaptic foci. The extracellular potentials usually vary in amplitude considerably when compared to intracellular EPSP potentials (Usherwood, 1972).

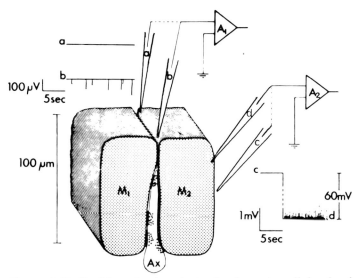

Figure 4-63. Position of microelectrodes for extracellular focal recording (a, b) and for intracellular recording (c, d) from the retractor unguis muscle. M$_{1,2}$, muscle cells; Ax, axon nerve terminal with synaptic sites. From Usherwood (1974).

f. Maggot Intersegmental Muscle

As mentioned above, synaptic foci are sometimes difficult to locate on the ordinary skeletal muscle fibers of adult insects. This is particularly true of fibrillar flight muscles of Diptera whose neuromuscular junctions are invaginated deeply within each muscle fiber and are relatively inaccessible (Neal, 1975a,b). On the other hand the metathoracic retractor unguis muscles from the adult fleshfly, *Sarcophaga bullata*, were found to have many neuromuscular junctions on the surface of the muscle (Neal, 1975b).

A relatively new nerve–muscle preparation from the third instar larvae of the blowfly, *Calliphora erythrocephala*, has been described recently (Hardie, 1976; Irving et al., 1976; Hardie and Irving, 1977; Crossley, 1965, 1968). The body wall muscles of blowfly larvae, *Lucilia, Phormia*, or *Calliphora*, are relatively large, about 2.0×0.4 mm for longitudinal ventrolateral muscles of *Calliphora* (muscles 6A and 7A of Crossley, 1965). They are dually innervated by two axons which may classified on the basis of two distinct excitatory postsynaptic potentials (EPSPs), a large fast EPSP and a small slow EPSP. Only the dorsal inner surface of the muscle is innervated (Hardie, 1976).

The most common characteristic of these muscles is the lack of a transverse invagination of the sarcolemma. In effect the muscle is a single "giant" fiber with a single tubular sheet of sarcolemma. Moreover, the tubular sheet of membrane may be peeled off in a special ionic solution containing 22 mM Na$_2$SO$_4$, 4 mM MgSO$_4 \cdot$ 7H$_2$O, 2 mM EDTA, 20 mM Tris buffer (pH 7.0), and 330 mM sucrose. The sheath of membrane contains intact nerve endings and is intact except for openings at both ends marking the exit hole from which the cytoplasm was removed. Each sheath weighs about 1.7 μg dry with about 0.64 μg of this being protein, and it is calculated that motor nerve terminals occupy 0.5 to 1% of the volume of the isolated sheath (Hardie and Irving, 1977).

The intersegmental muscle of the blowfly is certainly a unique insect muscle. It may be far more responsive to perfused glutamate than any other insect muscle because of the exposed and accessible synaptic terminals (Irving et al., 1976). Although studies on synaptic neurochemistry and neurophysiology are only just beginning for this muscle, the suggestion has already been made that the so-called "slow" and "fast" excitatory neuromuscular transmissions are mediated by different neurotransmitters (Irving and Miller, 1979, in press, *J. Comp. Physiol.*). This represents a rather profound new idea and suggests that the innervation of arthropod muscle is far from being described completely.

Acknowledgments

A book of this sort, concerned as it is with how problems of recording have been solved, necessarily depends heavily on new ideas from individuals. Where published, these ideas are referred to in the references; however, many of the circuits, electrodes, or other devices described here are the unpublished or otherwise unavailable work of certain colleagues. They are listed here with the pertinent figure numbers in parentheses: M. Véró (1-1), R. Chang (1-2), R. K. Josephson and P. L. Donaldson (1-5, 1-7, 1-10, and 4-4), R. Evans (1-14), C. Roemmele (1-21), P. Buchan (1-22, 2-21, and 2-22), V. Salgado (1-26b and 4-18), H. Spencer (1-26a, 1-27, and 1-28), M. D. R. Jones (2-14), R. Pence (2-17 and 2-18), E. W. Hamilton (2-20), G. Stange (2-24 and 2-25), M. Luff and L. Molyneux (2-26, and 2-27), E. Edney (2-38), D. Gammon (3-4), M. Adams (3-10, 3-11, and 3-13), R. Hustert (3-17), P. Miller (4-1), D. Graham (4-3), D. Sandeman (4-9), J. Barnes (4-10), B. Cook (4-11, 4-32, 4-33, and 4-34), T. Nagai (4-15, 4-38, and 4-48), R. House (4-39 and 4-40), M. Lafon-Cazal (4-41), D. Shankland (4-43 and 4-44), M. Schroeder (4-45 and 4-46), and D. Rees (4-58, 4-59, 4-60, and 4-61).

To those mentioned above and to everyone else who gave permission to reproduce original work and who cooperated, I am indebted.

My own unpublished work which is drawn upon here was supported by grants from the Rockefeller Foundation, by EPA Grant No. R 804345 awarded to my colleague T. R. Fukuto, and by NIH Grant Nos. ES 00814 and NS 09357.

Expert technical assistance was provided by Steve Manweiler, Dennis McFarland, and Frederick Tsao who redrew figures. The editors of Springer-Verlag, New York, assisted with the figures by lettering where needed. Mrs. Steve Miller typed the manuscript twice under difficult conditions and helped organize figures and references.

I am grateful also to P. Capstick, L. Stones, and C. Potter of Wellcome Research Laboratories who generously provided space and support during a sabbatical leave at Wellcome Research Laboratories, Berkhamsted, England and to Professor D. R. Newth who acted as host during another sabbatical leave at the Zoology Department, University of Glasgow. Both of these visits were enormously useful in gathering information and insight.

It would be quite involved to thank everyone who provided information for this book or who suffered my visits. However, I am especially grateful to Dr. Tom Piek of the University of Amsterdam who rescued the early manuscript from the Dover-Hook van Holland ferry for me at one point and forwarded it to Konstanz, Germany. The bulk of the information comes from the literature and follow-up letters for more detail, and it was supplemented by visits to laboratories in Europe which were undertaken in 1973, 1974, and 1976 and numerous visits to laboratories in the United States.

There is a considerable amount of information left out of this book on neurophysiological techniques. Most conspicuously, much modern information of central nervous recording methods is missing. This was necessary in the end for many reasons. However, depending on the response, perhaps future revisions might include more information on numerous subjects or, even better, contributions from other authors. Certainly the present work should be considered as a limited individual attempt to describe the enormous contributions of many scientists and technicians from all over the world.

Note Added in Proof

A few new devices have appeared recently: Buchan and Sattelle (*Physiological Entomology* **4:** 103–9, 1979) reported a radar-Doppler actograph; Luff et al. (*Ibid.* pp. 147–53) reported an ultrasonic movement detector. Hamilton has published his generalized circuitry for actographs in Volume 6 no. 4 of the *Journal of Electrophysiological Techniques* (pp. 3–8). Véró and Miller (*Medical & Biological Engineering & Computing,* 1979) reported a new force transducer that measures in the median nanogram range and can measure the force of Malpighian tubule muscle contraction.

The neuropeptide proctolin may be purchased from: Peninsula Laboratories, Inc., P. O. Box 1111, San Carlos, California 94070; or Serva (Feinbiochemica · Heidelberg) 28 Tec Street, Hicksville, New York 11801.

Muijser (*Experientia* **35:** 912–13, 1979) has reported a circuit for a microelectrode amplifier with infinite resistance and current injection capabilities. These devices are limited by the inherent noise level in the operational amplifier used in the input stage.

Recent work mentions inhibitory innervation of the retractor unguis muscle in *Schistocerca gregaria* (Clark, et al *Nature* **280:** 679–82, 1979) which is a new finding.

References

Aidley, D.J. (1971). *The Physiology of Excitable Cells*. Cambridge: Cambridge University Press.

Albrecht, F.O. (1953). *The Anatomy of the Migratory Locust*. London: Athlone Press.

Alexandrowicz, J.S. (1926). The innervation of the heart of the cockroach (*Periplanta orientalis*). *J. Comp. Neurol.* **41**: 291–310.

Andrew, B.L. (1972). *Experimental Physiology*. 9th ed. Baltimore: Williams and Wilkins Co.

Andrieu, A.-J. (1968). Nouvelles conceptions de l'actographie Appliquée aux Études des Invertebres. *Ann. Epiphyties* **19**:483–499.

Arnold, A.J. (1965). An electro-pneumatic microelectrode puller. *J. Sci. Instrum.* **42**:723–724.

Arnold, L.K. (1968). *Introduction to Plastics*. Ames, Iowa: University Press.

Asher, W.C., and Merritt, T. (1970). High density display and recording of pulse activity. *Ann. Entomol. Soc. Am.* **63**:1472–1473.

Aston, R.J., and White, A.F. (1974). Isolation and purification of the diuretic hormone from *Rhodnius prolixus*. *J. Insect Physiol.* **20**:1673–1682.

Atwood, H.L., and Parnas, I. (1966). Simple transistorized nerve stimulators. *Can. J. Physiol. Pharmacol.* **44**:487–491.

Atwood, H.L., Smyth, T., and Johnston, H.S. (1969). Neuromuscular synapses in the cockroach extensor tibiae muscle. *J. Insect Physiol.* **15**:529–536.

Axenfeld, D. (1911). Locomozione ae ea degli insetti. *Bull. Acad. Med. Roma* **37**:123–136.

Backlund, H.O., and Ekeroot, S. (1950). An actograph for small terrestrial animals. *Oikos* **2**:213–216.

Ball, H.J. (1971). The receptor site for photic entrainment of circadian activity rhythms in the cockroach *Periplaneta americana*. *Ann. Entomol. Soc. Am.* **64**:1010–1015.

Ball, H.J. (1972). A system for recording activity of small insects. *J. Econ. Entomol.* **65**:129–132.

Barrett, J., and Graubard, K. (1970). Fluorescent staining of cut motoneurons *in vivo* with beveled micropipettes. *Brain Res.* **18**:565–568.

Barrett, J.N., and Whitlock, D.G. (1973). Technique for beveling glass microelectrodes. In: *Intracellular Staining in Neurobiology.* S.B. Kater and C. Nicholson (Eds.). New York, Berlin: Springer-Verlag.

Barton-Browne, L., and Evans, D.R. (1960). Locomotor activity of the blowfly as a function of feeding and starvation. *J. Insect Physiol.* **4**:27–37.

Bastian, J. (1972). Neuro-muscular mechanisms controlling a flight maneuver in the honeybee. *J. Comp. Physiol.* **77**:126–140.

Batth, S.S. (1972). Pesticide effectiveness and biorhythm in the housefly. *J. Econ. Entomol.* **65**:1191–1193.

Bauer, H.C. (1976). Effects of photoperiod and temperature on the cholinesterase activity in the ganglia of *Schistocerca gregaria. J. Insect Physiol.* **22**:683–688.

Beard, R.L. (1952). The toxicology of *Habrobracon* venom: A study of a natural insecticide. *Conn. Agr. Expt. Sta. Bull.* 562: 27 pages.

Beard, R.L. (1960). Electrographic recording of foregut activity in larvae of *Galleria mellonella. Ann. Entomol. Soc. Am.* **53**:346–351.

Bělař, K. (1929). Beitrage zur Kausalanalyse der Mitose. II. Untersuchungen an den Spermatocyten von *Chorthippus (Stenobothrus) lineatus* Panz. *Wilhelm Roux Arch. Ento-Mech. Org.* **118**:359–484.

Belton, P., and Brown, B.E. (1969). The electrical activity of cockroach visceral muscle fibers. *Comp. Biochem. Physiol.* **28**:853–863.

Bennett, R.R., Buchan, P.B., and Treherne, J.E. (1975). Sodium and lithium movements and axonal function in cockroach nerve cords. *J. Exp. Biol.* **62**:231–241.

Bentley, D.R. (1969). Intracellular activity in cricket neurons during the generation of behaviour patterns. *J. Insect Physiol.* **15**:677–699.

Bentley, D.R., and Kutsch, W. (1966). The neuromuscular mechanism of stridulation in crickets (*Orthoptera: Gryllidae). J. Exp. Biol.* **45**:151–164.

Beránek, R., and Miller, P.L. (1968). The action of iontophoretically applied glutamate on insect muscle fibres. *J. Exp. Biol.* **49**:83–93.

Berridge, M.J., and Patel, N.G. (1968). Insect salivary glands: Stimulation of fluid secretion by 5-hydroxytryptamine and adenosine—3', 5'—monophosphate. *Science* **162**:462–463.

Berridge, M.J., and Prince, W.T. (1972). The role of cyclic AMP and calcium in hormone action. *Adv. Insect Physiol.* **9**:1–49.

Bettini, S., d'Ajello, V., and Maroli, M. (1973). Cartap activity on the cockroach nervous and neuromuscular transmission. *Pestic. Biochem. Physiol.* **3**:199–205.

Biston, J., and Sillans, D. (1976). Étude à l'aide d'un composant photosensible des variations du rythme circulatoire des chenilles de *Bombyx mori,* soumises à une anesthésie à l'éther. *Entomol. Exp. Appl.* **19**:255–262.

Blakeslee, T.R. (1975). *Digital Design with Standard MSI and LSI.* New York: John Wiley.

Bland, K.P., and House, C.R. (1971). Function of the salivary glands of the cockroach, *Nauphoeta cinerea. J. Insect Physiol.* **17:**2069–2084.

Bland, K.P., House, C.R., Ginsborg, B.L., and Laszlo, I. (1973). Catecholamine transmitter for salivary secretion in the cockroach. *Nature, New Biol.* **244:**26–27.

Boettiger, E.G., and Furshpan, E. (1952). The recording of flight movements in insects. *Science* **116:**60–61.

Bowser-Riley, F., and House, C.R. (1976). The actions of some putative neurotransmitters on the cockroach salivary gland. *J. Exp. Biol.* **64:**665–676.

Bradford, S., and Ramsey, W. (1949). Analysis of mosquito tissues for sodium and potassium and development of a physiological salt solution. *Fed. Proc.* **8:**15–16.

Bradshaw, C.M., Roberts, M.H.T., and Szabadi, E. (1973a). Kinetics of the release of noradrenaline from micropipettes: Interaction between ejecting and retaining currents. *Brit. J. Pharmacol.* **49:**667–677.

Bradshaw, C.M., Roberts, M.H.T., and Szabadi, E. (1973b). A theoretical model of ion-movements in micropipettes occurring during the course of microelectrophoresis experiments. *Br. J. Pharmacol.* **47:**653 P.

Brady, J. (1967a). Haemocytes and the measurement of potassium in insect blood. *Nature (London)* **215:**96–97.

Brady, J. (1967b). Control of the circadian rhythm of activity in the cockroach. I. The role of the corpora cardiaca, brain and stress. *J. Exp. Biol.* **47:**153–163.

Brady, J. (1967c). Control of the circadian rhythm of activity in the cockroach. II. The role of the sub-oesophageal ganglion and ventral nerve cord. *J. Exp. Biol.* **47:**165–178.

Brady, J. (1969). How are insect circadian rhythms controlled? *Nature (London)* **223:**781–784.

Brady, J. (1972). Spontaneous, circadian components of tsetse fly activity. *J. Insect Physiol.* **18:**471–484.

Brandt, P.W., Chappell, E., and Lewell, B.R. (1976). A robust transducer suitable for measuring forces 1 μH. *J. Physiol.* **258:**43–44.

Brian, M.V. (1947). On the ecology of beetles of the genus *Agriotes* with special reference to *A. obscurus. J. Anim. Ecol.* **16:**210–224.

Brookes, N., and Werman, R. (1973). The cooperativity of gamma-aminobutyric acid action on the membrane of locust muscle fibers. *Mol. Pharmacol.* **9:**571–579.

Brown, B.E. (1965). Pharmacologically active constituents of the cockroach corpus cardiacum: Resolution and some characteristics. *Gen. Comp. Endocrinol.* **5:**387–401.

Brown, B.E. (1967). Neuromuscular transmitter substance in insect visceral muscle. *Science* **155:**595–597.

Brown, B.E. (1975). Proctolin: A peptide transmitter candidate in insects. *Life Sci.* **17:**1241–1252.

Brown, B.E., and Nagai, T. (1969). Insect visceral muscle: Neural relations of the proctodeal muscles of the cockroach. *J. Insect Physiol.* **15:**1767–1783.

Brown, K.T., and Flaming, D.G. (1974). Beveling of fine micropipette electrodes by a rapid precision method. *Science* **185:**693–695.

Brown, P.B., Maxfield, B.W., and Moraff, H. (1973). *Electronics for Neurobiologists*. Cambridge, Mass.: MIT Press.

Brown, R.H.J. (1959). A simple activity recorder. *J. Insect Physiol.* **3**:125–126.

Brown, R.H.J., and Unwin, D.M. (1961). An activity recording system using infra-red detection. *J. Insect Physiol.* **7**:203–209.

Bruce, V.G., and Minis, D.H. (1969). Circadian clock action spectrum in a photoperiodic moth. *Science* **165**:583–585.

Burkhardt, D. (1964). Color discrimination in insects. *Adv. Insect Physiol.* **2**:131–173.

Burns, M.D. (1973). The control of walking in orthoptera. I. Leg movements in normal walking. *J. Exp. Biol.* **58**:45–58.

Burrows, M. (1974). Modes of activation of motoneurons controlling ventilatory movements of the locust abdomen. *Phil. Trans. R. Soc. London B* **269**:29–48.

Burrows, M. (1975). Co-ordinating interneurons of the locust which convey two patterns of motor commands. *J. Exp. Biol.* **63**:713–753.

Burrows, M., and Hoyle, G. (1973). Neural mechanisms underlying behavior in the locust *Schistocerca gregaria*. III. Topography of limb motorneurons in the metathoracic ganglion. *J. Neurobiol.* **4**:167–186.

Burrows, M., and Siegler, M.V.S. (1976). Transmision without spikes between locust interneurones and motoneurones. *Nature (London)* **262**:222–224.

Burt, P.E., and Goodchild, R.E. (1971). The site of action of pyrethrin I in the nervous system of the cockroach *Periplaneta americana*. *Entomol. Exp. Appl.* **14**:179–189.

Burt, P.E., and Goodchild, R.E. (1974). Knockdown by pyrethroids: Its role in the intoxication process. *Pestic. Sci.* **5**:625–633.

Burton, R.F. (1975). *Ringer Solutions and Physiological Salines*. Bristol: Scientechnica.

Bushland, R.C. (1975). Screwworm research and eradication. *Bull. Entomol. Soc. Am.* **21**:23–26.

Busvine, J.R. (1957). *A Critical Review of the Techniques for Testing Insecticides*. London: Commonwealth Institute of Entomology.

Callec, J.-J. (1974). Synaptic transmission in the central nervous system of insects. In: *Insect Neurobiology*. J.E. Treherne (Ed.). Amsterdam: North-Holland.

Callec, J.-J., and Sattelle, D.B. (1973). A simple technique for monitoring the synaptic actions of pharmacological agents. *J. Exp. Biol.* **59**:725–738.

Cambridge, G.W., and Haines, J. (1959). A new versatile transducer system. *J. Physiol.* **149**:2–3

Camhi, J.M. (1970). Yaw-correcting postural changes in locusts. *J. Exp. Biol.* **52**:519–531.

Campan, R. (1972). Light-induced heartbeat disturbances: Comparative study in *Calliphora vomitoria* (Linnaeus 1758) (Diptera) and *Nemobius sylvestris* (Bosc 1792) (Orthoptera). *Monitore Zool. Ital.* (N.S.) **6**:269–289.

Carbonell, C.S. (1947). The thoracic muscles of the cockroach, *Periplaneta americana* L. *Smithsonian Misc. Coll.* **107**:1–23.

Cavallin, C., Persson, B., and Ulfstrand, S. (1972). Total activity of caged birds; a new recording method. *Oikos* **23**:140–141.

Chamberlain, S., Kerkut, G.A., and Venning, H.B. (1966). Negative capacitance solid state microelectrode input circuit for use in neurophysiology. *Life Sci.* **5**:743–745.

Chambers, D.L. (1975). Quality in mass-produced insects: Definition and evaluation. In: *Controlling Fruit Flies by the Sterile-Insect Technique*, pp. 19–32. International Atomic Energy Agency, Vienna.

Chang, J.J. (1975). A new technique for beveling the tips of glass capillary micropipettes and microelectrodes. *Comp. Biochem. Physiol.* **52A**:567–570.

Chapman, K.M. (1965). Campaniform sensillae on the tactile spines of the legs of the cockroach. *J. Exp. Biol.* **42**:191–203.

Chapman, K.M., and Nichols, T.R. (1969). Electrophysiological demonstration that cockroach tibial tactile spines have separate sensory axons. *J. Insect Physiol.* **15**:2103–2115.

Chapman, K.M., and Pankhurst, J.H. (1967). Conduction velocities and their temperature coefficients in sensory nerve fibres of cockroach legs. *J. Exp. Biol.* **46**:63–84.

Chowdhury, T.K. (1969). Extremely fine glass microelectrodes. *J. Sci. Instrum.* **2**:1087–1090.

Clark, R.B. (1966). *A Practical Course in Experimental Zoology*. London: Wiley.

Clarke, K.U., and Grenville, H. (1960). Nervous control of movements in the foregut in *Schistocerca gregaria* Forsk. *Nature (London)* **186**:98–99.

Clements, A.N. (1963). *Physiology of Mosquitoes*. Oxford: Pergamon Press.

Cloudsley-Thompson, J.L. (1955). The design of entomological actograph apparatus. *Entomologist* **88**:153–161.

Cloudsley-Thompson, J.L. (1961). *Rhythmic Activity in Animal Physiology and Behavior*. New York and London: Academic Press.

Cochrane, D.G., Elder, H., and Usherwood, P.N.R. (1972). Physiology and ultrastructure of phasic and tonic skeletal muscle fibres in the locust, *Schistoceria gregaria*. *J. Cell Sci.* **10**:419–441.

Coggshall, J.C., Boschek, C.B., and Buchner, S.M. (1973). Preliminary investigations on a pair of giant fibers in the central nervous system of dipteran flies. *Z. Naturforsch.* **28**:783–784.

Collett, J.I. (1976). Some features of the regulation of the free amino acids in adult *Calliphora erythrocephala*. *J. Insect Physiol.* **22**:1395–1403.

Collins, C., and Miller, T. (1977). Studies on the action of biogenic amines on cockroach heart. *J. Exp. Biol.* **67**:1–15.

Coombes, J.S. (1965). Some methods of reducing interference caused by stimulation artifacts. In: *Studies in Physiology*. D.R. Curtis and A.K. McIntyre (Eds.). New York:Springer-Verlag.

Cook, B.J. (1967). An investigation of Factor S, a neuromuscular excitatory substance from insects and crustacea. *Biol. Bull. (Woods Hole, Mass.)* **133**:526–538.

Cook, B.J., and Holman, G.M. (1975a). Neural control of *Leucophaea* hindgut. *J. Comp. Physiol.* **84**:95–118.

Cook, B.J., and Holman, G.M. (1975b). Hindgut muscle activity in *Leucophaea maderae*. *J. Insect physiol.* **21**:1187–1192.

Cook, B.J., and Reinecke, J.P. (1973). Visceral muscles and myogenic activity in the hindgut of the cockroach, *Leucophaea maderae*. *J. Comp. Physiol.* **84**:95–118.

Cook, B.J., Eraker, J., and Anderson, G.R. (1969). The effect of various biogenic amines on the activity of the foregut of the cockroach, *Blaberus giganteus*. *J. Insect Physiol.* **15**:445–455.

Cook, B.J., Long, J.B., and Owens, D. (1971). An inexpensive solid state amplifier for measuring bioelectric potentials. *Comp. Biochem. Physiol.* **40A**:385–390.

Cooper, R. (1956). Storage of silver chloride electrodes. *Electroencephalogr. Clin. Neurophysiol.* **8**:692.

Cooper, R.D., Umemoto, L., Friend, W.G., and Hartwick, W.B. (1971). Effects of drugs on the behaviour of the carpenter ant, *Camponotus herculeanus*. *Comp. Gen. Pharmacol.* **2**:120–123.

Corning, W.C., Feinstein, D.A., and Haight, J.R. (1965). Arthropod preparation for behavioral, electrophysiological, and biochemical studies. *Science* **148**:394–395.

Courtice, C.J. (1976). A current pump monitor for microiontophoresis circuits. *J. Physiol.* **258**:42–43.

Crescitelli, F., and Jahn, T.L. (1938). Electrical and mechanical aspects of the grasshopper cardiac cycle. *J. Cell. Comp. Physiol.* **11**:359–376.

Crossley, A.C. (1965). Transformation in the abdominal muscles of the blue blowfly, *Calliphora erythrocephala* (Meig.), during metamorphosis. *J. Embryol. Exp. Morphol.* **14**:89–110.

Crossley, A.C. (1968). The fine structure and mechanism of breakdown of larval intersegmental muscles in the blowfly, *Calliphora erythrocephala*. *J. Insect Physiol.* **14**:1389–1407.

Crowder, L.A., and Shankland, D.L. (1972a). Pharmacology of the malpighian tubule muscle of the American cockroach *Periplaneta americana*. *J. Insect Physiol.* **18**:929–936.

Crowder, L.A., and Shankland, D.L. (1972b). Structure of the malpighian tubule muscle of the American cockroach, *Periplaneta americana*. *Ann. Entomol. Soc. Am.* **65**:614–618.

Cull-Candy, S.G., Neal, H. and Usherwood, P.N.R. (1973). Action of black widow spider venom on an aminergic synapse. *Nature (London)* **241**:353–354.

Curtis, D.R. (1964). Microelectrophoresis. In: *Physical Techniques in Biological Research,* Vol. 5. W.L. Nastuk (Ed.). New York: Academic Press.

Cymborowski, B. (1969). The effect of light on the circadian rhythm of locomotor activity of the house cricket *Acheta domestica* L. *Zool. Pol.* **19**:85–95.

Cymborowski, B. (1972). Construction of an apparatus for recording the locomotor activity of insects. *Exp. Physiol. Biochem.* **5**:229–255.

Cymborowski, B. (1973). Control of the circadian rhythm of locomotor activity in the house cricket. *J. Insect Physiol.* **19**:1423–1440.

Dade, H.A. (1962). *Anatomy and Dissection of the Honeybee.* London: Bee Research Association.

Davey, K.G. (1964). The control of visceral muscles in insects. *Adv. Insect Physiol.* **2**:219–245.

David, J., and Rougier, M. (1972). *Physiology of the explanted* dorsal vessel of insects. In: *Invertebrate Tissue Culture.* C. Vago (Ed.). New York: Academic Press.

Davson, H. (1964). *A Textbook of General Physiology,* 3rd ed. Boston: Little, Brown and Co.

Delcomyn, F. (1971). Computer aided analysis of a locomotor leg reflex in the cockroach *Periplaneta americana. Z. Vgl. Physiol.* **74:**427–445.

Delcomyn, F. (1973). Motor activity during walking in the cockroach *Periplaneta americana.* II. Tethered walking. *J. Exp. Biol.* **59:**643–654.

Delcomyn, F. (1974). A simple system for suction electrodes. *J. Electrophysiol. Tech.* **3:**22–25.

Delcomyn, F. (1976). An approach to the study of neural activity during behavior in insects. *J. Insect Physiol.* **22:**1223–1227.

Delcomyn, F., and Usherwood, P.N.R. (1973). Motor activity during walking in the cockroach, *Periplaneta americana.* I. Free walking. *J. Exp. Biol.* **59:**629–642.

Demerec, M. (1950). *Biology of Drosophila.* New York: Wiley.

Dethier, V.G. (1974). Sensory input and the inconstant fly. In: *Experimental Analysis of Insect Behavior.* L. Barton-Browne (Ed.). Berlin: Springer-Verlag.

Dethier, V.G., and Gelperin, A. (1967). Hyperphagia in the blowfly. *J. Exp. Biol.* **47:**191–200.

Dewhurst, D.J. (1976). *An Introduction to Biomedical Instrumentation.* 2nd ed. Oxford, New York: Pergamon Press.

de Wilde, J. (1947). Contribution to the physiology of the heart of insects, with special reference to the alarm muscles. *Arch. Neerl. Physiol.* **28:**530–542.

Dibley, G.C., and Lewis, T. (1972). An ant counter and its use in the field. *Entomol. Exp. Appl.* **15:**499–508.

Dichter, M.A. (1973). Intracellular single unit recording. In: *Bioelectric Recording Techniques—Part A: Cellular Processes and Brain Potentials,* R.F. Thompson and M.M. Patterson (Eds.). pp. 3–21. New York: Academic Press.

Dierichs, R. (1972). Elektronenmikroskopische Untersuchungen des ventralen Diaphragms von *Locusta migratoria* und der langsamen Kontraktionswelle nach Fixation durch Gefriersubstitution. *Z. Zellforsch.* **126:**402–420.

Dobkin, R.C. (1973). Bridge amplifier. *Elec. Eng. Times,* Monday, July 16 issue.

Donley, C.S. (1975). A device for fast and foolproof filling of glass micropipettes. *J. Electrophysiol. Tech.* **4:**15–17.

Dowson, R.J., and Usherwood, P.N.R. (1972). The effect of low concentrations of L-glutamate and L-aspartate on transmitter release at the locust excitatory nerve–muscle synapse. *J. Physiol.* **229:**13–14.

Dreisig, H. (1971). Diurnal activity in the dusky cockroach, *Ectobius lapponicus* L. (Blattodea). *Entomol. Scand.* **2:**132–138.

Dreisig, H., and Nielsen, E.T. (1971). Circadian rhythm of locomotion and its temperature dependence in *Blattella germanica. J. Exp. Biol.* **54:**187–198.

Drenth, D. (1969). Some aspects of the collection and the action of the venom of *Microbracon hebetor* (Say). *Acta Physiol. Pharmacol. Neerl.* **15:**100.

Drenth, D. (1974a). Susceptibility of different species of insects to an extract of the venom gland of the wasp *Microbracon hebetor* (Say). *Toxicon* **12:**189–192.

Drenth, D. (1974b). Stability of *Microbracon hebetor* (Say) venom preparations. *Toxicon* **12**:541–542.

Dreyer, F., and Peper, K. (1974a). Iontophoretic application of acetylcholine: Advantages of high resistance micropipettes in connection with an electronic current pump. *Pflügers Arch.* **348**:263–272.

Dreyer, F., and Peper, K. (1974b). The acetylcholine sensitivity in the vicinity of the neuromuscular junction of the frog. *Pflügers Arch.* **348**:273–286.

Dudel, J. (1975). Potentiation and desensitization after glutamate induced post-synaptic currents at the crayfish neuromuscular junction. *Pflügers Arch.* **356**:317–327.

Duling, B.R. and Berne, R.M. (1969). A rapid, small-volume method for filling micropipettes. *J. Appl. Physiol.* **26**:837.

Dutky, S.R., Schechter, M.S., and Sullivan, W.N. (1963). Monitoring electrophysiological and locomotor activity of insects to detect biological rhythms. In: *Biotelemetry*. L.E. Slater (Ed.). New York: Pergamon Press.

Duwez, Y. (1936). Electrocardiogram of *Dytiscus*. *C.R. Soc. Biol. Paris*. **122**:84–87.

Duwez, Y. (1938). L'Automatisme cardiaque chez lè dytique. *Arch. Int. Physiol.* **46**:389–403.

Eastham, L. (1925). Peristalsis in the Malpighian tubules of Diptera. *Quart. J. Microsc. Sci.* **69**:385–398.

Edney, E.B. (1937). A study of spontaneous locomotor activity in *Locusta migratoria migratorioides* (R. and F.) by the actograph method. *Bull. Entomol. Res.* **28**:243–278.

Edney, E.B. (1966). Absorption of water vapour from unsaturated air by *Arenivaga* sp. (Polyphagidae, Dictyoptera). *Comp. Biochem. Physiol.* **19**:387–462.

Edney, E.B., Haynes, S., and Gibo, D. (1974). Distribution and activity of the desert cockroach *Arenivaga investigata* (Polyphagidae) in relation to microclimate. *Ecology* **55**:420–427.

Edwards, D.K. (1958). Two quantitative methods for measuring insect activity. *Can. Entomol.* **90**:612–616.

Edwards, D.K. (1960). A method for continuous determiniation of displacement activity in a group of flying insects. *Can. J. Zool.* **38**:1021–1025.

Edwards, D.K. (1964). Activity rhythms of Lepidopterous defoliators. *Can. J. Zool.* **42**:923–937.

Elsner, N. (1968a). Muskelaktivitat und Lauterzeugung bei einer Felcheuschrecke. *Zool. Anz.* **31**:592–601.

Elsner, N. (1968b). Die neuromuskulären Grundlagen des Werbeverhaltens der Roten Keulenheuschrecke *Gomphocerippus rufus* (L.). *Z. Vgl. Physiol.* **60**:308–350.

Elsner, N. (1970). The records of stridulatory movements in the grasshopper *Chorthippus mollis* using Hall-generators. *Z. Vgl. Physiol.* **68**:417–428.

Elsner, N. (1974a). Neuroethology of sound production in Gomphocerine grasshoppers (*Orthoptera: Acrididae*). Part I. Song patterns and stridulatory movements. *J. Comp. Physiol.* **88**:67–102.

Elsner, N. (1974b). Neural economy: Bifunctional muscles and common central pattern elements in leg and wing stridulation of the grasshopper *Stenobothrus rubicundus*, Germ. (*Orthoptera: Acrididae*). *J. Comp. Physiol.* **89**:227–236.

Elsner, N., and Huber, F. (1969). Die Organisation des Werbegesanges der Heuschrecke *Gomphocerippus rufus* in Abhängigkeit von zentralen und periphere Bedingungen. *Z. Vgl. Physiol.* **65**:389–423.

Elsner, N., and Popov, A.V. (1978). Neuroethology of acoustic communication. *Adv. Insect Physiol.* **13**:229–355.

Engelmann, F. (1963). Die Innervation der Genital-und Post-genitalsegmente bie Weibchen der Schabe *Leucophaea maderae. Zool. Jahrb. Anat.* **81**:1–63.

Engelmann, F. (1968). Feeding and crop emptying in the cockroach *Leucophaea maderae. J. Insect Physiol.* **14**:1525–1532.

Engelmann, F. (1970). *The Physiology of Insect Reproduction.* Oxford: Pergamon Press.

Ephrusse, B., and Beadle, G.W. (1936). A technique of transplantation for *Drosophila. Am. Nat.* **70**:218–225.

Erber, J. (1975). Turning behavior related to optical stimuli in *Tenebrio molitor. J. Insect Physiol.* **21**:1575–1580.

Evans, P.D. (1972). The free amino acid pool of the haemocytes of *Carcinus maenas* (L.). *J. Exp. Biol.* **56**:501–507.

Evans, P.D. (1975). The uptake of L-glutamate by the central nervous system of the cockroach, *Periplaneta americana. J. Exp. Biol.* **62**:55–67.

Ewing, A.W., and Manning, A. (1966). Some aspects of the efferent control of walking in three cockroach species. *J. Insect Physiol.* **12**:1115–1118.

Faeder, I.L.R. (1968). Neuromuscular transmission in insects. Ph.D. dissertation, Cornell University, Ithaca, New York.

Feder, W. (1968). Bioelectrodes. *Ann. N.Y. Acad. Sci.* **148**:1–287.

Fein, H. (1966). Passing current through recording glass micropipette electrodes. *IEEE Trans. Biomed. Eng.* **13**:211–212.

Ferris, C.O. (1975). *Introduction to Bioelectrodes.* New York: Plenum Press.

Finger, F.W. (1972). Measuring behavioral activity. In: *Methods in Psychobiology,* Vol. 2. R.D. Myers (Ed.). London: Academic Press.

Flattum, R.F., and Shankland, D.L. (1971). Acetylcholine receptors and the diphasic action of nicotine in the American cockroach, *Periplaneta americana* (L.). *Comp. Gen. Pharmacol.* **2**:159–167.

Flattum, R.F., Watkinson, I.A., and Crowder, L.A. (1973). The effect of insect "autoneurotoxin" in *Periplaneta americana* (L.) and *Schistocerca gregaria* (Forskal) Malpighian tubules. *Pestic. Biochem. Physiol.* **3**:237–242.

Florey, E., and Kriebel, M.E. (1966). A new suction electrode system, *Comp. Biochem. Physiol.* **18**:175–178.

Florkin, M., and Jeuniaux, C. (1974). Hemolymph: Composition. In: *The Physiology of Insecta,* Vol. V, 2nd ed. M. Rockstein (Ed.). New York: Academic Press.

Fondacaro, J.D., and Butz, A. (1970). Circadian rhythm of locomotor activity and susceptibility to methyl parathion of adult *Tenebrio molitor* (Coleoptera: Tenebrionidae). *Ann. Entomol. Soc. Am.* **63**:952–955.

Fourtner, C.R. (1976). Morphological properties of fast and slow skeletal muscles in the cockroach. *Am. Zool.* **16**:173.

Frank, K.D., and Zimmerman, W.F. (1969). Action spectra for phase shifts of a circadian rhythm in *Drosophila. Science* **163**:688.

Fredman, S.M., and Steinhardt, R.A. (1973). Mechanism of inhibitory action by salts in the feeding behaviour of the blowfly, *Phormia regina. J. Insect Physiol.* **19**:781–790.

274   References

Freeman, M.A. (1966). The effect of drugs on the alimentary canal of the African migratory locust *Locusta migratoria*. *Comp. Biochem. Physiol.* **17**:755–764.

Gammon, D.W. (1977). Nervous effects of toxins on an intact insect: A method. *Pestic. Biochem. Physiol.* **7**:1–7.

Garoutte, B., and Lie, K.H. (1972). Tungsten microneedles: A simple method of production. *Electroencephalogr. Clin. Neurophysiol.* **33**:425–426.

Geddes, L.A. (1970). Bioelectric impedance. *Science* **167**:1761.

Geddes, L.A. (1972). *Electrodes and the Measurement of Bioelectric Events*. New York: Wiley.

Geddes, L.A., and Baker, L.E. (1975). *Principles of Applied Biomedical Instrumentation*, 2nd ed. New York: Wiley.

Geiger, G. (1974). Optomotor responses of the fly *Musca domestica* to transient stimuli of edges and stripes. *Kybernetik* **16**:37–43.

Gelperin, A. (1967). Stretch receptors in the foregut of the blowfly. *Science* **157**:208–210.

Gelperin, A. (1971). Regulation of feeding. *Annu. Rev. Entomol.* **16**:365–377.

Gelperin, A. (1972). Neural control systems underlying insect feeding behavior. *Am. Zool.* **12**:489–496.

Getting, P.A. (1971). The sensory control of motor output in the fly proboscis extension. *Z. Vgl. Physiol.* **74**:103–120.

Getting, P.A., and Steinhardt, R.A. (1972). The interaction of external and internal receptors in the feeding behaviour of the blowfly, *Phormia regina* Meigen. *J. Insect Physiol.* **18**:1673–1681.

Gewecke, M. (1975). The influence of the air-current sense organs on the flight behavior of *Locusta migratoria*. *J. Comp. Physiol.* **102**:79–95.

Ginsborg, B.L., House, C.R., and Silinsky, E.M. (1974). Conductance changes associated with the secretory potential in the cockroach salivary gland. *J. Physiol.* **236**:723–731.

Ginsborg, B.L., House, C.R., and Silinsky, E.M. (1976). On the actions of compounds related to dopamine at a neurosecretory synapse. *Brit. J. Pharmacol.* **57**:133–140.

Girardie, A., and Lafon-Cazal, M. (1972). Contrôle endocrine des contractions de l'oviducte isolé de *Locusta migratoria migratorioides* (R. et F.). *C.R. Acad. Sci. Paris*, D **274**:2208–2210.

Glanzman, D.L., and Glanzman, F. (1973). Pipettes. *Carrier* **1**:1–3. (Published by David Kopf Instruments.)

Glaser, R.W. (1925). Hydrogen ion concentration in the blood of insects. *J. Gen. Physiol.* **7**:599–602.

Godden, D.H. (1973). A re-examination of circadian rhythmicity in *Carausius morosus*. *J. Insect Physiol.* **19**:1377–1386.

Godden, D.H., and Goldsmith, T.H. (1972). Photoinhibition of arousal in the stick insect *Carausius*. *Z. Vgl. Physiol* **76**:135–145.

Goetz, K.G. (1972). Processing of cues from the moving environment in the *Drosophila* navigation system. In: *Information Processing in the Visual Systems of Arthropods*, pp. 255–263. R. Wehner (Ed.). Berlin: Springer-Verlag.

Goldsmith, T.H., and Bernard, G.D. (1974). The visual system of insects. In: *The Physiology of Insecta*, Vol. 2, 2nd ed., pp. 165–272. M. Rockstein (Ed.). New York: Academic Press.

Goryshin, N.I., Braun, E.A., and Tyshchenko, G.F. (1973). Multichannel device for studying some daily rhythms in insects. *Vestn. Leningr. Univ. Ser. B* **28**:21–25.

Graham-Smith, G.S. (1934). The alimentary canal of *Calliphora erythrocephala* L., with special reference to its musculature and to the proventriculus, rectal valve and rectal papillae. *Parasitology* **26**:176–248.

Green, J.D. (1958). A simple microelectrode for recording from the central nervous system. *Nature (London)* **182**:962.

Green, G.W. (1964). The control of spontaneous locomotor activity in *Phormia regina* Meigen—I. Locomotor activity patterns in intact flies. *J. Insect Physiol.* **10**:711–726.

Green, G.W., and Anderson, D.C. (1961). A simple and inexpensive apparatus for photographing events of pre-set intervals. *Can. Entomol.* **93**:741–745.

Gregory, G.E. (1974). Neuroanatomy of the mesothoracic ganglion of the cockroach *Periplaneta americana* (L). I. The roots of the peripheral nerves. *Philos. Trans. R. Soc. Lond. Ser. B* **267**:421–465.

Griffiths, J.G., and Smyth, T. (1973). Action of black widow spider venom at insect neuromuscular junctions. *Toxicon* **11**:369–374.

Griffiths, J.G., and Tauber, O.E. (1943). Effects of pH and of various concentrations of sodium, potassium, calcium chlorides in muscular activity of the isolated crop of *Periplaneta americana* (Orthoptera). *J. Gen. Physiol.* **26**:541–558.

Grobbelaar, J.H., Morrison, G.J., Baart, E.E., and Moran, V.C. (1967). A versatile, highly sensitive activity recorder for insects. *J. Insect Physiol.* **13**:1843–1848.

Grundfest, H., Sengstaken, R.W., Oettinger, W.H., and Curry, R.W. (1950). Stainless steel micro-needle electrodes made by electrolytic pointing. *Rev. Sci. Instrum.* **21**:360–361.

Guthrie, D.M. (1962). Control of the ventral diaphragm in an insect. *Nature (London)* **196**:1010–1012.

Guthrie, D.M. and Tindall, A.R. (1968). *Biology of the Cockroach*. London: Arnold.

Hagadorn, H.H. (1974). The control of vitellogenesis in the mosquito, *Aedes aegypti*. *Am. Zool.* **14**:1207–1217.

Hamilton, H.L. (1939). The action of acetylcholine, atropine, and nicotine on the heart of the grasshopper, *Melanoplus differentialis*. *J. Cell. Comp. Physiol.* **13**:91–104.

Hamilton, E.W. (1976). Air puff sensor for insect electrophysiological studies. *J. Electrophysiol. Tech.* **5**:40–41.

Hammond, J.H. (1954). An actograph for small insects. *J. Sci. Instrum.* **31**:43–44.

Hardeland, R., and Stange, G. (1971). Einflüsse von Geschlecht und Alter auf die locomotische Aktivität von *Drosophila*. *J. Insect Physiol.* **17**:427–434.

Hardeland, R., and Stange, G. (1973). Comparative studies on the circadian rhythms of locomotor activity of 40 *Drosophila* species. *J. Interdiscipl. Cycle Res.* **4**:353–359.

Hardie, J. (1976). Motor innervation of the supercontracting longitudinal ventrolateral muscles of the blowfly larvae. *J. Insect Physiol.* **22**:661–668.

Hardie, J. and Irving, S.N. (1977). A preparation enriched with insect somatic, excitatory, neuromuscular terminals. *Brain Res.*, **120**:138–140.

Hays, R.D. (1953). Determination of a physiological saline solution for *Aedes aegypti* (L.). *J. Econ. Entomol.* **46**:624–627.

Hearney, E.S. (1975). Neuropile organization of the metathoracic ganglion of *Periplaneta americana* (L.) (Dictyoptera: Blattidae). *Int. J. Insect Morphol. Embryol.* **4**:265–272.

Heide, G. (1975). Properties of a motor output system involved in the optomotor response in flies. *Biol. Cybernet.* **20**:99–112.

Heinrich, B. (1970). Nervous control of the heart during thoracic temperature regulation in a sphinx moth. *Science* **169**:606–607.

Heinrich, B. (1971). Temperature regulation of the sphinx moth, *Manduca sexta*. II. Regulation of heat loss by control of blood circulation, *J. Exp. Biol.* **54**:153–166.

Heinrich, B. (1976). Heat exchange in relation to blood flow between thorax and abdomen in bumblebees. *J. Exp. Biol.* **64**:561–585.

Heit, M., Sauer, J.R., and Mills, R.R. (1973). The effects of high concentrations of sodium in the drinking medium of the American cockroach, *Periplaneta americana* (L.). *Comp. Biochem. Physiol.* **45A**:363–370.

Hewitt, C.G. (1914). *The Housefly*. Ithaca: Cornell University Press.

Highnam, K.C. (1962). Neurosecretory control of ovarian development in *Schistocerca gregaria*. *Q. J. Microsc. Sci.* **103**:57–72.

Highnam, K.C. (1964). Endocrine relationships in insect reproduction. *Symp. R. Entomol. Soc. London* **2**:26–42.

Hilliard, S.D., and Butz, A. (1969). Daily fluctuations in the concentrations of total sugar and uric acid in the hemolymph of *Periplaneta americana*. *Ann. Entomol. Soc. Am.* **62**:71–74.

Hobson, A.D. (1928). The effect of electrolytes on the muscle of the foregut of *Dytiscus marginalis* with special reference to the action of potassium. *J. Exp. Biol.* **5**:385–393.

Hocking, B. (1953). The intrinsic range and speed of flight of insects. *Trans. R. Entomol. Soc. London* **104**:223–345.

Hodgson, E.S., and Roeder, K.D. (1956). Electrophysiological studies of arthropod chemoreception. I. General properties of the labellar chemoreceptors of Diptera. *J. Cell. Comp. Physiol.* **48**:51–75.

Hodgson, E.S., Lattvin, J.Y., and Roeder, K.D. (1955). The physiology of a primary chemoreceptor unit. *Science* **122**:417.

Hoenig, S.A., and Payne, F.L. (1973). *How to Build and Use Electronic Devices without Frustration, Panic, Mountains of Money, or an Engineering Degree.* Boston: Little, Brown & Co.

Holden, R.E.D., and Sattelle, D.B. (1972). A multiway nonreturn valve for use in physiological experiments. *J. Physiol.* **226**:2–3P.

Holloway, R.L., and Smith, J.W. (1975). Locomotor activity of adult lesser cornstalk borer. *Ann. Entomol. Soc. Am.* **68**:885–887.

Holman, G.M., and Cook, B.J. (1970). Pharmacological properties of excitatory neuromuscular transmission in the hindgut of the cockroach, *Leucophaea maderae*. *J. Insect Physiol.* **16**:1891–1907.

Holman, G.M., and Cook, B.J. (1972). Isolation, partial purification, and characterization of a peptide which stimulates the hindgut of the cockroach, *Leucophaea maderae* (Fabr.). *Biol. Bull. (Woods Hole, Mass.)* **142**:446–460.

Holman, G.M., and Marks, E.P. (1974). Synthesis, transport, and release of a neurohormone by cultured neuroendrocrine glands from the cockroach, *Leucophaea maderae. J. Insect Physiol.* **20**:479 – 485.

House, C.R. (1973). An electrophysiological study of neuroglandular transmission in the isolated salivary glands of the cockroach. *J. Exp. Biol.* **58**:29–43.

House, C.R. (1975). Intracellular recording of secretory potentials in a "mixed" salivary gland. *Experientia* **31**:904–906.

House, C.R., and Ginsborg, B.L. (1976). Actions of a dopamine analogue and a neuroleptic at a neuroglandular synapse. *Nature (London)* **261**:332–333.

House, C.R., Ginsborg, B.L., and Silinsky, E.M. (1973). Dopamine receptors in cockroach salivary gland cells. *Nature, New Biol.* **245**:63.

Hoyle, G. (1955a). The anatomy and innervation of locust skeletal muscle. *Proc. R. Soc. London B* **143**:281–292.

Hoyle, G. (1955b). Neuromuscular mechanisms of a locust skeletal muscle. *Proc. R. Soc. London B.* **143**:343–367.

Hoyle, G. (1964). Exploration of neural mechanisms underlying behavior in insects. In: *Neural Theory and Modelling.* R.F. Reiss (Ed.). Stanford: Stanford University Press.

Hoyle, G. (1966). Functioning of the inhibitory conditioning axon innervating insect muscles. *J. Exp. Biol.* **44**:429–454.

Hoyle, G. (1968). Slow and fast axon control of contraction in insects. *Exp. Physiol. Biochem.* **1**:287–297.

Hoyle, G. (1970). Cellular mechanisms underlying behavior—neuroethology. *Adv. Insect Physiol.* **7**:349–444.

Hoyle, G. (1974). Neural control of skeletal muscle. In: *The Physiology of Insecta,* Vol. IV, 2nd ed. M. Rockstein (Ed.). New York: Academic Press.

Hoyle, G. (1975). Identified neurons and the future of neuroethology. *J. Exp. Zool.* **194**:51–74.

Hoyle, G., and Burrows, M. (1973). Neural mechanisms underlying behavior in the locust *Schistocerca gregaria.* I. Physiology of identified motorneurons in the metathoracic ganglion. *J. Neurobiol.* **4**:3–41.

Hsaio, H.S. (1972). The attraction of moths (*Trichoplusia ni*) to infrared radiation. *J. Insect Physiol.* **18**:1705–1714.

Huber, F. (1960). Untersuchungen über die Funktion des Zentralnervensystems und insbesondere des Gehirns bei der Fortbewegung und der Lauterzeugung der Grillen. *Z. Vgl. Physiol.* **44**:60–132.

Huber, F. (1974). Neural integration (central nervous system). In: *The Physiology of Insecta,* 2nd ed. M. Rockstein (Ed.). New York: Academic Press.

Hue, B., Pelhate, M., Callec, J.J., and Chanelet, J. (1975). Synaptic action of 4 amino pyridine in abdominal ganglion of the cockroach *Periplaneta americana. Soc. Biol. Fil.* **169**:876–883.

Huettel, M.D. (1975). Monitoring the quality of laboratory-reared insects. *J. N. Y. Entomol. Soc.* **83**:276.

Hughes, P.R., and Pitman, G.B. (1970). A method of observing and recording the flight behaviour of tethered Bark Beetles in response to chemical messengers. *Contrib. Boyce Thompson Inst.* **24**:329–336.

Hustert, R. (1974). Morphologic und Atmungsbewegungen des 5. Abdominalsegmentes von *Locusta migratoria migratorioides. Zool. Jahrb. Physiol.* **78:**157–174.

Hustert, R. (1975). Neuromuscular coordination and proprioceptive control of rhythmical abdominal ventilation in intact *Locusta migratoria migratorioides. J. Comp. Physiol.* **97:**159–179.

Huxley, A.F., and Simmons, R.M. (1968). A capacitance-gauge tension transducer. *J. Physiol.* **197:**12P.

Ikeda, K. (1977). Flight motor innervation of a fleshfly. In: *Identified Neurons and Behavior of Arthropods,* pp. 357–368. G. Hoyle (Ed.). New York: Plenum Press.

Ikeda, K., and Boettiger, E.G. (1965). Studies on the flight mechanism of insects. II. The innervation and electrical activity of the fibrillar muscles of the bumblebee, *Bombus. J. Insect Physiol.* **11:**779–789.

Ikeda, K., and Kaplan, W.D. (1970). Patterned neural activity of a mutant *Drosophila melanogaster. Proc. Natl. Acad. Sci. USA* **66:**765–772.

Ikeda, K., Hori, N., and Tsuruhara, T. (1976). Motor innervation of the dorsal longitudinal flight muscle of *Sarcophaga bullata. Am. Zool.* **16:**179.

Iles, J.F. (1972). Structure and synaptic activation of the fast coxal depressor motoneurone of the cockroach *Periplaneta americana. J. Exp. Biol.* **56:**647–656.

Irisawa, H., Irisawa, A.F., and Kadotani, T. (1956). Findings on the electrograms of the cicada's heart (*Cryptotympania japonensis* Kato). *Jap. J. Physiol.* **6:**150–161.

Irving, S.N., Osborne, M.P., and Wilson, R.G. (1976). Virtual absence of L-glutamate from the haemoplasm of arthropod blood. *Nature (London)* **263:**431–433.

Jahn, T.L., Crescitelli, F., and Taylor, A.B. (1937). The electrocardiogram of the grasshopper (*Melanoplus differentialis*). *J. Cell. Comp. Physiol.* **10:**439–460.

Johnson, B. (1966). Fine structure of the lateral cardiac nerves of the cockroach *Periplaneta americana* (L.). *J. Insect Physiol.* **12:**645–653.

Jones, J.C. (1960). The anatomy and rhythmical activities of the alimentary canal of *Anopheles* larvae. *Ann. Entomol. Soc. Am.* **53:**459–474.

Jones, J.C. (1977). *The Circulatory System of Insects.* Springfield: Charles C Thomas.

Jones, M.D.R. (1964). The automatic recording of mosquito activity. *J. Insect Physiol.* **10:**343–351.

Jones, M.D.R., and Reiter, P. (1975). Entrainment of the pupation and adult activity rhythms during development in the mosquito *Anopheles gambiae. Nature (London)* **254:**242–244.

Jones, M.D.R., Hill, M., and Hope, A.M. (1967). The circadian flight activity of the mosquito *Anopheles gambiae:* Phase setting by the light regime. *J. Exp. Biol.* **47:**503–511.

Jones, M.D.R., Cubbin, C.M., and Marsh, D. (1972). The circadian rhythm of flight activity of the mosquito *Anopheles gambiae:* The light-response rhythm. *J. Exp. Biol.* **57:**337–346.

Jones, R.H., and Potter, H.W. (1972). A six-position artificial feeding apparatus for *Culicoides variipennis. Mosquito News* **32:**520–527.

Josephson, R.K., and Halverson, R.C. (1971). High frequency muscles used in sound production by a katydid. 1. Organization of the motor system. *Biol. Bull. (Woods Hole, Mass.)* **141**:411–433.

Josephson, R.K., Stokes, D.R., and Chen, V. (1975). The neural control of contraction in a fast insect muscle. *J. Exp. Zool.* **193**:281–300.

Kaiser, W., and Bishop, L.G. (1970). Directionally selective motion detecting units in the optic lobe of the honeybee. *Z. Vgl. Physiol.* **67**:403–413.

Kammer, A.E. (1967). Muscle activity during flight in some large Lepidoptera. *J. Exp. Biol.* **47**:277–295.

Kammer, A.E. (1971). The motor output during turning flight in a hawkmoth, *Manduca sexta. J. Insect Physiol.* **17**:1073–1086.

Kanehisa, K. (1965). Effect of temperature on the isolated hindgut movement of American cockroach *Periplaneta americana* L. *Jap. J. Appl. Entomol. Zool.* **9**:301–302.

Kanehisa, K. (1966a). Pharmacological studies on the isolated hindgut movement of American cockroach, *Periplaneta americana* L. (Blattaria:Blattidae). I. Effects of cholinesters, bioactive amines, amino acids and several autonomic nerve agents. *Appl. Entomol. Zool.* **1**:83–93.

Kanehisa, K. (1966b). Pharmacological studies on the isolated hindgut movement of American cockroach, *Periplaneta americana* L. (Blattaria:Blattidae). II. Effect of Insecticides. *Appl. Entomol. Zool.* **1**:145–153.

Kashin, P. (1966). Electronic recording of the mosquito bite. *J. Insect Physiol.* **12**:281–286.

Kashin, P., and Arneson, B.E. (1969). An automated repellency assay system. II. A new electronic "bitometer-timer." *J. Econ. Entomol.* **62**:200–205.

Kashin, P., and Wakeley, H.G. (1965). An insect "bitometer." *Nature (London)* **208**:462–464.

Kaufman, W.R., and Davey, K.G. (1971). The pulsatile organ in the tibia of *Triatoma phyllosoma pallidipennis. Can. Entomol.* **103**:487–496.

Kay, R.H., and Coxon, R.V. (1956). Simultaneous recording of inspired oxygen concentration and peripheral tissue oxygenation. *Nature (London)* **177**:45–46.

Kelly, J.S., Simmonds, M.A., and Straughn, D.W. (1975). Microelectrode techniques. In: *Methods in Brain Research.* P.G. Bradley (Ed.). London: Wiley.

Kennedy, J.S., and Booth, C.O. (1963). Free flight of aphids in the laboratory. *J. Exp. Biol.* **40**:67–85.

Kerfoot, W.B. (1966). A photoelectric activity recorder for studies of insect behavior. *J. Kansas Entomol. Soc.* **39**:629–633.

Kerkut, G.A., and Walker, R.J. (1967). The effects of iontophoretic injection of L-glutamic acid and γ-aminobutyric acid on the miniature endplate potentials and contractures of the coxal muscles of the cockroach, *Periplaneta americana* L. *Comp. Biochem. Physiol.* **20**:999–1003.

Kerkut, G.A., Pitman, R.M., and Walker, R.J. (1969). Iontophoretic application of acetylcholine and GABA onto insect central neurones. *Comp. Biochem. Physiol.* **31**:611–633.

King, A.B.S. (1973). The actographic examination of flight activity of the cocoa mirid *Distantiella theomobroma* (Hemiptera:Miridae). *Entomol. Exp. Appl.* **16**:53–63.

Kishaba, A.N., Henneberry, T.J., Hancock, P.J., and Toba, H.H. (1967). Laboratory technique for studying flight of cabbage looper moths and the effects of age, sex, food, and Tepa on flight characteristics. *J. Econ. Entomol.* **60**:359–366.

Klemm, H. (1976). Histochemistry of putative transmitter substances in the insect brain. *Progr. Neurobiol.* **7**:99–169.

Klostermeyer, E.C., and Gerber, H.S. (1969). Nesting behavior of *Megachila rotundata* (Hymenoptera:Megachilidae) monitored with an event recorder. *Ann. Entomol. Soc. Am.* **62**:1321–1325.

Klostermeyer, E.C., Mech, S.J., and Rasmussen, W.B. (1973). Sex and weight of *Megachila rotundata* (Hymenoptera:Megachilidae) progeny associated with provision weights. *J. Kansas Entomol. Soc.* **46**:536–548.

Knight, M.R. (1962). Rhythmic activities of the alimentary canal of the black blowfly, *Phormia regina. Ann. Entomol. Soc. Am.* **55**:380–382.

Koch, U. (1978). A miniature movement detector applied to recording of wingbeat in *Locusta.* In: *Physiology of Locomotion-Biomechanics.* W. Nachtigall (Ed.). West Germany: Fischer-Verlag, in press.

Kogan, M. (1972). Electronic recordings of feeding motions and rhythms of corn earworm (*Heliothis zea*) larvae. *Israel J. Entomol.* **7**:49–58.

Kogan, M. (1973). Automatic recordings of masticatory motions of leaf-chewing insects. *Ann. Entomol. Soc. Am.* **66**:66–69.

Kogan, M., and Goedden, R.D. (1971). Feeding and host-selection behavior of *Lema trilineota doturaphila* larvae (Coleoptera: Chrysomelidae). *Ann. Entomol. Soc. Am.* **64**:1435–1448.

Koller, G. (1948). Rhythmische Bewegung and hormonale steuerung bei den Malpighischen Gefässen der Insekten. *Biol. Zentralbl.* **67**:201–211.

Kooistra, G. (1950). Contribution to the knowledge of the action of acetycholine in the intestine of *Periplaneta americana* L. *Physiol. Comp. Oecol.* **21**:75–80.

Kramer, E. (1976). The orientation of walking honeybees in odour fields with small concentration gradients. *Physiol. Entomol.* **1**:27–37.

Krijgsman, B.D., Dresden, D., and Berger, N.E. (1950). The action of rotenone and TEPP on the isolated heart of the cockroach. *Bull. Entomol. Res.* **41**:141–151.

Kripke, BR., and Ogden, T.E. (1974). A technique for beveling fine micropipettes. *Electroencephalogr. Clin. Neurophysiol.* **36**:323–326.

Krogh, A., and Weis-Fogh, T. (1952). A roundabout for studying sustained flight of locusts. *J. Exp. Biol.* **29**:211–219.

Kutsch, W. (1969). Neuromuskuläre Aktivität bei verschiedenen Verhaltensweisen von Drei Grillenarten. *Z. Vgl. Physiol.* **63**:335–378.

Kutsch, W., and Otto, D. (1972). Evidence for spontaneous song production independent of head ganglia in *Gryllus campestris* L. *J. Comp. Physiol.* **81**:115–119.

Kuznetsov, N.Ya. (1915). *Fauna of Russia and Adjacent Countries,* Vol. I. Lepidoptera, USSR.

Lang, F. (1972). Electrophysiology of the neurogenic heart of *Limulus. Exp. Physiol. Biochem.* **5**:127–154.

Lavallec, M., Schanne, O.F., and Hebert, N.C. (1969). *Glass Microelectrodes.* New York: Wiley.

Leppla, N.C., and Spangler, H.G. (1971). A flight cage actograph for recording circadian periodicity of Pink Bollworm moths. *Ann. Entomol. Soc. Am.* **64**:1431–1434.

Lewis, A.F. (1967). Constant voltage double pulse generators for use in electrophysiology. *Electr. Eng.* **39**:252–258.

Lion, K.S. (1964). Transducers. In: *Physical Techniques in Biological Research*, Vol. V. W.L. Nastuk (Ed.). New York: Academic Press.

Lipton, G.R., and Sutherland, D.J. (1970a). Activity rhythms in the American cockroach, *Periplaneta americana. J. Insect Physiol.* **16**:1555–1566.

Lipton, G.R., and Sutherland, D.J. (1970b). Feeding rhythms in the American cockroach, *Periplaneta americana. J. Insect Physiol.* **16**:1757–1767.

Löfqvist, J., and Stenram, H. (1965). An electronic actograph for automatic registration of animal locomotion. *Acta Univ. Lund.* **11**:1–9.

Loher, W. (1974). Circadian control of spermatophore formation in the cricket *Teleogryllus commodus* Walker. *J. Insect Physiol.* **20**:1155–1172.

Loher, W., and Chandrashekaran, M.K. (1970). Circadian rhythmicity in the oviposition of the grasshopper *Chorthippus curtipennis. J. Insect Physiol.* **16**:1677–1688.

Lowne, B.T. (1890-1892). *The Anatomy, Physiology, Morphology, and Development of the Blow-fly. (Calliphora erythrocephala).* London: R.H. Porter.

Ludwig, D., Tracey, K.M., and Burns, M.L. (1957). Ratios of ions required to maintain the heartbeat of the American cockroach *Periplaneta americana* Linnaeus. *Ann. Entomol. Soc. Am.* **50**:244–246.

Luff, M.L., and Molyneux, L. (1970). Insect movement detector using operational amplifiers. *J. Physics E* **3**:939–941.

Lutz, F.E. (1932). Experiments with Orthoptera concerning diurnal rhythm. *Am. Museum Novitates* No. 550:1–6.

Lux, H.D., Schubert, P., Kreutzberg., G.W., and Globus, A. (1970). Excitation and axonal flow: Autoradiographic study on motoneurons intracellularly injected with a ^{3}H-amino acid. *Exp. Brain Res.* **10**:197–204.

Macauley, E.D.M. (1972). A simple insect flight recorder. *Entomol Exp. Appl.* **15**:252–254.

Macauley, E.D.M. (1974). Lipid storage in the pre-imago and young adult *Plusia gamma. Entomol. Exp. Appl.* **17**:53–60.

Machan, R., and Himstedt, W. (1970). Ein neuer kapazitiv arbeitender Aktograph. *Oecologia* **4**:211–217.

Maddrell, S.H.P. (1971). The mechanisms of insect excretory systems. *Adv. Insect Physiol.* **8**:200–331.

Maddrell, S.H.P., and Klunsuwan, S. (1973). Fluid secretion by *in vitro* preparations of the Malpighian tubules of the desert locust *Schistocerca gregaria. J. Insect Physiol.* **19**:1369–1376.

Marks, E.P., Holman, G.M., and Borg, T.K. (1973). Synthesis and storage of a neurohormone in insect brains *in vitro. J. Insect Physiol.* **19**:471–477.

Marrelli, J.D., and Hsiao, H.S. (1976). Miniature angle transducer for marine arthropods. *Comp. Biochem. Physiol.* **54A**:121–123.

Martin, A.R., Wickelgren, W.O., and Béranek, R. (1970). Effects of iontophoretically applied drugs on spinal interneurone of the lamprey. *J. Physiol.* **207**:653–665.

McCaffery, A.R. (1976). Effects of electrocoagulation of cerebral neurosecretory cells and implantation of corpora allata on oöcyte development in *Locusta migratoria*. *J. Insect Physiol.* **22**:1081–1092.

McCann, F.V. (1969). The insect heart as a model for electrophysiological studies. *Exp. Physiol. Biochem.* **2**:59–88.

McDonald, T.J. (1975). Neuromuscular pharmacology of insects. *Annu. Rev. Entomol.* **20**:151–166.

McIndoo, N.E. (1939). Segmental blood vessels of the American cockroach (*Periplaneta americana* L.). *J. Morphol.* **65**:323–348.

McIndoo, N.E. (1945). Innervation of insect hearts. *J. Comp. Neurol.* **83**:141–155

McLean, D.L., and Kinsey, M.G. (1964). A technique for electronically recording aphid feeding and salvation. *Nature (London)* **202**:1358–1359.

McLean, D.L., and Kinsey, M.G. (1965). Identification of electrically recorded curve patterns associated with aphid salivation and ingestion. *Nature (London)* **205**:1130–1131.

McLean, D.L., and Kinsey, M.G. (1967). Probing behavior of the pea aphid *Acyrthosiphon pisum*. I. Definitive correlation of electronically recorded waveforms with aphid probing activities. *Ann. Entomol. Soc. Am.* **60**:400–406.

McLean, D.L., and Kinsey, M.G. (1968). Probing behavior of the pea aphid, *Acyrthosiphon pisum*. II. Comparison of salivation and ingestion in host and non-host plant leaves. *Ann. Entomol. Soc. Am.* **61**:730–739.

McLean, D.L., and Kinsey, M.G. (1969). Probing behavior of the pea aphid, *Acyrthosiphon pisum*. IV. Effects of starvation on certain probing activities. *Ann. Entomol. Soc. Am.* **62**:993–994.

McLean, D.L., and Weigt, W.A. (1968). An electronic measuring system to record aphid salivation and ingestion. *Ann. Entomol. Soc. Am.* **61**:180–185.

McLennan, H., and Wheal, H.V. (1976). The interaction of glutamic and aspartic acids with excitatory amino acid receptors in the mammalian central nervous system. *Can. J. Physiol. Pharmacol.* **54**:70–72.

Medioni, J. (1964). Une nouvelle technique d'enregistrement actographique applicable de enregistrement actographique applicable à la Drosophile et à des animaux plus petits. *C.R. Soc. Biol., Paris* **158**:2185.

Merrill, E.G. (1972a). A simple light-coupled stimulus isolator. *J. Physiol.* **222**:120–121P.

Merrill, E.G. (1972b). A log display for neurone firing rate vs. time. *J. Physiol.* **224**:2–3P.

Merrill, E.G. (1972c). A versatile high input impedance differential amplifier. *J. Physiol.* **224**:1–2P.

Merrill, E.G., and Ainsworth, A. (1972). Glass-coated platinum-plated tungsten microelectrodes. *Med. Biol. Eng.* **10**:662–672.

Meyer, H. (1976). The different actions of chloride and potassium on postsynaptic inhibition of an isolated neurone. *J. Exp. Biol.* **64**:477–487.

Meyers, J., and Miller, T. (1969). Starvation-induced activity in cockroach Malpighian tubules. *Ann. Entomol. Soc. Am.* **62**:725–729.

Mill, P.J., and Pickard, R.S. (1972). Anal valve movement and normal ventilation in Aeshnid dragonfly larvae. *J. Exp. Biol.* **56**:537–543.

Miller, J.A., Eschle, J.L., and Berry, I.L. (1969). Patterns of flight activity in livestock insects. 1. Preliminary testing of a system for recording flight activity of the stable fly. *Ann. Entomol. Soc. Am.* **62**:1046–1050.

Miller, P.L. (1971a). Rhythmic activity in the insect nervous system. I. Ventilatory coupling of a mantid spiracle. *J. Exp. Biol.* **54**:587–597.

Miller, P.L. (1971b). Rhythmic activity in the insect nervous system. II. Sensory and electrical stimulation of ventilation in a mantid. *J. Exp. Biol.* **54**:599–607.

Miller, P.L. (1971c). Rhythmic activity in the insect nervous system. Thoracic ventilation in non-flying beetles. *J. Insect Physiol.* **17**:395–405.

Miller, P.L. (1973). Spatial and temporal change in the coupling of cockroach spiracles to ventilation. *J. Exp. Biol.* **58**:137–148.

Miller, P.L. (1974a). Rhythmic activities and the insect nervous system. In: *Experimental Analysis of Insect Behavior.* L. Barton-Browne (Ed.). Berlin: Springer-Verlag.

Miller, P.L. (1974b). Respiration—aerial gas transport. In: *The Physiology of Insecta,* Vol. VI, 2nd ed. M. Rockstein (Ed.). New York: Academic Press.

Miller, T. (1968). Role of cardiac neurons in the cockroach heartbeat. *J. Insect Physiol.* **14**:1265–1275.

Miller, T. (1969). Initiation of activity in the cockroach heart. *Experientia Suppl.* **15**:206–218.

Miller, T. (1973). Regulation of heartbeat of the American cockroach. In: *Neurobiology of Invertebrates.* J. Salánki (Ed.). Budapest: Akadémiai Kiadó.

Miller, T. (1974a). Electrophysiology of the insect heart. In: *The Physiology of Insecta,* Vol. 4, 2nd ed. M. Rockstein (Ed.). New York: Academic Press.

Miller, T.A. (1974b). Visceral muscle. In: *Insect Muscle.* P.N.R. Usherwood (Ed.). London: Academic Press.

Miller. T. (1975). Neurosecretion and the control of visceral organs in insects. *Annu. Rev. Entomol.* **20**:133–149.

Miller, T. (1976). Distinguishing between carbamate and organophosphate insecticide poisoning in house flies by symptomology. *Pestic. Biochem. Physiol.* **6**:307–319.

Miller, T., and Adams, M.E. (1974). Ultrastructure and electrical properties of the hyperneural muscle of *Periplaneta americana. J. Insect Physiol.* **20**:1925–1936.

Miller, T., and James, J. (1976). Chemical sensitivity of the hyperneural nerve muscle preparation of the American cockroach. *J. Insect Physiol.* **22**:981–988.

Miller, T., and Metcalf, R.L. (1968). A simple device for insect mechanocardiogram. *Ann. Entomol. Soc. Am.* **61**:1618–1620.

Miller, T., and Rees, D. (1973). Excitatory transmission in insect neuromuscular systems. *Am. Zool.* **13**:299–313.

Miller, T., and Usherwood, P.N.R. (1971). Studies of cardioregulation in the cockroach *Periplaneta americana. J. Exp. Biol.* **54**:329–348.

Miller, T., Bruner, L.J., and Fukuto, T.R. (1971). The effect of light, temperature, and DDT poisoning in housefly locomotion and flight muscle activity. *Pestic. Biochem. Physiol.* **1**:483–491.

Mitchell, B.K. (1976). Physiology of an ATP receptor in lobellar sensilla of the tsetse fly *Glossina morsitans* Wester. (Diptera:Glossinidae). *J. Exp. Biol.* **65**:259–271.

Möhl, B. (1972). The control of foregut movement by the stomatogastric nervous system in the European house cricket *Acheta domesticus* L. *J. Comp. Physiol.* **80**:1–28.

Moorhouse, J.E., Fosbrooke, I.H.M., and Kennedy, J.S. (1978). 'Paradoxical' driving of walking activity in insects. *J. Exp. Biol.* **72**:1–16.

Moreau, R. and Lavenseau, L. (1975). Rôle des organes pulsatiles thoraciques et du cœur pendant l'émergence et l'expansion des ailes des Lépidopteres. *J. Insect Physiol.* **21**:1531–1534.

Morita, H., and Yamashita, S. (1959). Generator potential of insect chemoreceptor. *Science* **130**:922.

Mulloney, B. (1969). Interneurons in the central nervous system of flies and the start of flight. *Z. Vgl. Physiol.* **64**:243–253.

Mulloney, B. (1970). Organization of flight motoneurons of diptera. **33**:86–95.

Mulloney, B. (1976). Control of flight and related behaviour by the central nervous systems of insects. In: *Insect Flight.* R.C. Rainey (Ed.). New York: Wiley.

Mulloney, B. and Selverston, A. (1972). Antidromic action potentials fail to demonstrate known interactions between neurons. *Science* **177**:69–72.

Nachtigall, W., and Wilson, D.M. (1967). Neuromuscular control of dipteran flight. *J. Exp. Biol.* **47**:77–97.

Nagai, T. (1970). Insect visceral muscle. Responses of the proctodeal muscles to mechanical stretch. *J. Insect Physiol.* **16**:437–448.

Nagai, T. (1972). Insect visceral muscle. Ionic dependence of electrical potentials in the proctodeal muscle fibres. *J. Insect Physiol.* **18**:2299–2318.

Nagai, T. (1973). Insect visceral muscle. Excitation and conduction in the proctodeal muscles. *J. Insect Physiol.* **19**:1753–1764.

Nagai, T., and Brown, B.E. (1969). Insect visceral muscle. Electrical potentials and contraction in fibres of the cockroach proctodeum. *J. Insect Physiol.* **15**:2151–2167.

Nagai, T., and Graham, G.T. (1974). Insect visceral muscle. Fine structure of the proctodeal muscle fibres. *J. Insect Physiol.* **20**:1999–2013

Narahashi, T. (1963). The properties of insect axons. *Adv. Insect Physiol.* **1**:175–256.

Narahashi, T. (1971). Effects of insecticides on excitable tissues. *Adv. Insect Physiol.* **8**:1–93.

Narahashi, T. (1976). Effects of insecticides on nervous conduction and synaptic transmission, In: *Insecticide Biochemistry and Physiology.* C.F. Wilkinson (Ed.). New York: Plenum.

Nayar, K.K. (1958). Studies on the neurosecretory system of *Iphita limbata* Stal. *Proc. Ind. Acad. Sci. B* **47**:233–251.

Naynert, M. (1968). Methode zur kontinuierlichen Registrierung der Atembewegungenvam freibeweglichen Krebs (*Potamobius astacus* Leach). *Experientia* **24**:1289–1290.

Neal, H. (1975a). Effects of L-glutamate and other drugs on some membrane properties of muscle fibres of Diptera. *J. Insect Physiol.* **21**:1771–1778

Neal, H. (1975b). Neuromuscular junctions and L-glutamate-sensitive sites in the fleshfly, *Sarcophaga bullata*. *J. Insect Physiol.* **21**:1945–1951.

Neville, A.C. (1963). Motor unit distribution of the locust dorsolongitudinal muscle. *J. Exp. Biol.* **40**:123–136.

Normann, T.C. (1972). Heart activity and its control in the adult blowfly, *Calliphora erythrocephala. J. Insect Physiol.* **18**:1793–1810.

Nüesch, H. (1953). The morphology of the thorax of *Telea polyphemus* (Lepidoptera). I. Skeleton and muscles. *J. Morphol.* **93**:589–610.

Nyboer, J. (1959). *Electrical Impedance Plethysmography.* Springfield: Charles C Thomas

Obara, Y. (1975). Mating behavior of the cabbage white butterfly, *Pieris rapae crucivora.* VI. Electrophysiological decision of muscle functions in wing and abdomen movements and muscle output patterns during flight. *J. Comp. Physiol.* **102**:189–200.

Odland, G.C., and Jones, J.C. (1975). Contractions of the hindgut of adult *Aedes aegypti* with special reference to the development of a physiological saline. *Ann. Entomol. Soc. Am.* **68**:613–616.

Ogden, T.E., Citron, M.C., and Pierantoni, R. (1978). The jet stream microbeveler: an inexpensive way to bevel ultrafine glass micropipettes. *Science* **201**:469–470.

Ögren, S.O. (1970). Motor activity measure with a new activity meter. *Science Tools* **17**:65–66.

Olsen, R.W., Ban, M., and Miller, T. (1976). Studies on the neuropharmacological activity of bicuculline and related compounds. *Brain Res.* **102**:283–299.

Omura, Y. (1970). Relationship between transmembrane action potentials of single cardiac cells and their corresponding surface electrograms *in vivo* and *in vitro,* and related electromechanical phenomena, *Trans. NY. Acad. Sci., Ser. II* **32**:874–910.

Ornberg, R.L., Smyth, T., and Benton, A.W. (1976). Isolation of a neurotoxin with a presynaptic action from the venom of the black window spider (*Latrodectus mactans,* Fabr.). *Toxicon* **14**:329–333.

O'Shea, M., Rowell, C.H.F., and Williams, J.L.D. (1974). The anatomy of a locust visual interneuron; the descending contralateral movement detector. *J. Exp. Biol.* **60**:1–12.

Otto, D. (1967). Untersuchungen zur nervösen kontrolle des Grillengesanges. *Zool. Anz. Suppl.* **31**:585–592.

Otto, D. (1971). Untersuchungen zur zentrolnervösen kontrolle der Lauterzengung von Grillen. *Z. Vgl. Physiol.* **74**:227–271.

Palka, J. (1969). Discrimination between movements of eye and object by visual interneurons of crickets. *J. Exp. Biol.* **50**:723–732.

Palm, N.B. (1946). Studies in the peristalsis of Malpighian tubules of insects. *Acta Univ. Lund.* **42**:1–39.

Parnas, I., and Dagan, D. (1971). Functional organizations of giant axons in the central nervous systems of insects: New aspects. *Adv. Insect Physiol.* **8**:95–144.

Parnas, I., Spira, M.E., Werman, R., and Bergman, F. (1969). Nonhomogeneous conduction in giant axons of the nerve cord of *Periplaneta americana. J. Exp. Biol.* **50**:635–649.

Parnas, I., Avgar, D., and Shulov, A. (1970). Physiological effects of venom of *Leiurus quinquestriatus* on neuromuscular systems of locust and crab. *Toxicon* **8**:67–79.

Pearson, K.G., (1972). Central programming and reflex control of walking in the cockroach. *J. Exp. Biol.* **56**:173–193.

Pearson, K.G. (1973). Function of peripheral inhibitory axons in insects. *Am. Zool.* **13**:321–330.

Pearson, K.G. (1976). The control of walking. *Sci. Am.* **235**:72–86.

Pearson, K.G., and Bergman, S.J. (1969). Common inhibitory motoneurons in insects. *J. Exp. Biol.* **50**:445–471.

Pearson, K.G., and Fourtner, C.R. (1973). Identification of the common inhibitory motoneurons in the metathoracic ganglion of the cockroach. *Can. J. Zool.* **51**:859–866.

Pearson, K.G., and Fourtner, C.R. (1975). Nonspiking interneurons in walking system of the cockroach. *J. Neurophysiol.* **38**:33–52.

Pearson, K.G., and Iles, J.F. (1971). Innervation of coxal depressor muscles in the cockroach *Periplaneta americana*. *J. Exp. Biol.* **54**:215–232.

Pearson, K.G., Stein, R.B., and Malhotra, S.K. (1970). Properties of action potentials from insect motor nerve fibres. *J. Exp. Biol.* **53**:299–316.

Pearson, K.G., Wong, R.K.S., and Fourtner, C.R. (1976). Connexions between hair-plate afferents and motoneurons in the cockroach leg. *J. Exp. Biol.* **64**:251–266.

Pence, R.J., Chambers, R.D., and Viray, M.S. (1963). "Psychogenetics stress" and autointoxication in the honeybee. *Nature (London)* **200**:930–932.

Pence, R.J., Viroy, M.S., Ebeling, W., and Reierson, D.A. (1975). Honey-bee abdomen assays of hemolymph from stressed and externally poisoned American cockroaches. *Pestic. Biochem. Physiol.* **5**:90–100

Perumpral, J.V., Earp, U.F., and Stanley, J.M. (1972). Strain gauge transducer to study the response of insects to physical stimuli. *Trans. Am. Soc. Agr. Engr.* **15**:785–787.

Pesson, P., and Girish, G.-K. (1968). Sensibilité des divers stades de développement de *Sitophilus zeamosis* Mote (-*S. oryzae* L.) aux radiations ionisantes étude des stades endogés par radiographié et enregistrement actographique. *Ann. Épiphyt.* **19**:513–531.

Pesson, P., and Ozer, M. (1968). Utilisation d'un octographe a détecteur électroacoustique pour l'élude des insectes des grain. *Ann. Épiphyt.* **19**:501–512.

Pesson, P., Ramade, F., and Riviere, J.L., and Gruson, H. (1971). Application d'une méthode actographique à l'étude de l'intoxication de la mouche domestique par divers insecticides de contact. *Entomol. Exp. Appl.* **14**:467–479.

Phaneuf, C., Langlois, J.-M., Roussel, J.-P., and Corrivault, G.-W. (1973). Procédé optique pour l'étude de l'activité cardiaque chez l'insecte. *C.R. Acad. Sci. Paris, D,* **276**:2899–2902.

Phipps, C.G., and Lucchina, G.G. (1964). Physiologic electrode tablet. U.S. Pat. Nos. 3, 137, 291.

Phipps, J. (1963). Laboratory observations on the activity of Acridoidea. *J. Insect Physiol.* **9**:531–543.

Pichon, Y. (1970). Ionic content of haemolymph in the cockroach, *Periplaneta americana*. *J. Exp. Biol.* **53**:195–209.

Pichon, Y. (1974). Axonal conduction in insects. In: *Insect Neurobiology.* J.E. Treherne (Ed.). Amsterdam: North Holland.

Pichon, Y., and Callec, J.J. (1970). Further studies on synaptic transmission in insects. I. External recording of synaptic potentials in a single giant axon of the cockroach, *Periplaneta americana*. *J. Exp. Biol.* **52:**257–265.

Pickard, R.S., and Mill, P.J. (1972). Ventilatory muscle activity in intact preparations of Aeshnid dragonfly larvae. *J. Exp. Biol.* **56:**527–536.

Pickard, R.S., and Mill, P.J. (1974). Ventilatory movements of the abdomen and bronchial apparatus in dragonfly larvae (Odonata: Anisoptera). *J. Zool.* **174:**23–40.

Pickard, R.S., and Mill, P.J. (1975). Ventilatory muscle activity in restrained and free-swimming dragonfly larvae (Odonata: Anisoptera). *J. Comp. Physiol.* **96:**37–52.

Pickard, R.S., and Welberry, T.R. (1976). Printed circuit microelectrodes and their application to honeybee brain. *J. Exp. Biol.* **63:**39–44.

Piek, T. (1966). Site of action of venom of *Microbracon hebetor* Say (Braconidae, Hymenoptera). *J. Insect Physiol.* **12:**561–568.

Piek, T., and Engels, E. (1969). Action of the venom of *Microbracon hebetor* (Say) on larvae and adults of *Philosomia cynthia* Hübn. *Comp. Biochem. Physiol.* **28:**603–618.

Piek, T., and Mantel, P. (1970a). The effect of the venom of *Microbracon hebetor* (Say) on the hyperpolarizing potentials in a skeletal muscle of *Philosamia cynthia* Hübn. *Comp. Gen. Pharmacol.* **1:**87–92.

Piek, T., and Mantel, P. (1970b). A study of the different types of action potentials and miniature potentials in insect muscles. *Comp. Biochem. Physiol.* **34:**935–951.

Piek, T., Mantel, P., and Engels, E. (1971). Neuromuscular block in insects caused by the venom on the digger wasp *Philanthus triangulum* F. *Comp. Gen. Pharmacol.* **2:**317–331.

Piek, T., Spanjer, W., Njio, K.D., Veenendaal, R.L., and Mantel, P. (1974). Paralysis caused by the venom of the wasp *Microbracon gelechiae*. *J. Insect Physiol.* **20:**2307–2320.

Pilcher, D.E.M. (1971). Stimulation of movements of Malpighian tubules of *Carausius* by pharmacologically active substances and tissue extracts. *J. Insect Physiol.* **17:**463–470.

Pipa, R.L., and Cook, E.F. (1959). Studies on the hexapod nervous system. I. The peripheral distribution of the thoracic nerves of the adult cockroach, *Periplaneta americana*. *Ann. Entomol. Soc. Am.* **52:**695–710.

Pittendrigh, C.S., Eichorn, J.H., Minis, D.H., and Bruce, V.G. (1970). Circadian systems. VI. The photoperiodic time meausurement in *Pectinophora gossypiella*. *Proc. Natl. Acad. Sci. USA* **66:**758–764.

Portier, P. (1949). *Encyclopedic Entomologique*, Vol. XXII, Ser. A. P. Luhevalier (Ed.).

Powell, J.A., Esch, H., and Craig, G.B. (1966). Electronic recording of mosquito activity. *Entomol. Exp. Appl.* **9:**385–394.

Prince, W.T., and Berridge, M.J. (1973). The role of calcium in the action of 5-hydroxytryptamine and cyclic AMP on salivary glands. *J. Exp. Biol.* **58:**367–384.

Pringle, T.N.S. (1938). Proprioception in insects. *J. Exp. Biol.* **15:**101–113.

Pumphrey, R.J., and Rawdon-Smith, A.F. (1937). Synaptic transmission of nerve impulses through the last abdominal ganglion of the cockroach. *Proc. R. Soc. London Ser. B.* **122**:106–118.

Quennic, Y., and Campan, M. (1972). Heartbeat frequency variations in the moth. *Mamestra brassicae* during ontogeny. *J. Insect Physiol.* **18**:1739–1744.

Quennic, Y., and Campan, M. (1975). Influence de la maturite sexuelle sur l'activité et la reactivité cardiaques de *Calliphora vomitoria* (Diptera, Calliphoridae). *J. Insect Physiol* **70**:457–466.

Rathmayer, W. (1962). Paralysis caused by the digger wasp *Philanthus. Nature (London)* **196**:1148–1151.

Rees, D. (1974). The spontaneous release of transmitter from insect nerve terminals as predicted by the negative binomial theorem. *J. Physiol.* **236**:129–142.

Reichardt, W. (1973). Musterinduzierte Flugorientierung. Verhaltens-Versuche an der Fiege *Musca domestica. Naturwissenschaften* **60**:122–138.

Reinecke, J.P., Cook, B.J., and Adams T.S. (1973). Larval hindgut of *Manduca sexta* (L.). (Lepidoptera:Sphingidae). *Int. J. Insect Morphol. Embryol.* **2**:277–290.

Rice, M.J., and Finlayson, L.H. (1972). Response of blowfly cibarial pump receptors to sinusoidal stimulation. *J. Insect Physiol.* **18**:841–846.

Richards, A.G. (1963). Effect of temperature on heart-beat frequency in the cockroach, *Periplaneta americana. J. Insect Physiol.* **9**:597–606.

Richter, K. (1967). Untersuchungen zum Wirkungsmechanismus von Neurohormon D am Herzen von *Periplaneta americana. Zool. Jahrb. Physiol* **73**:261–275.

Richter, K. (1973). Struktur und Funktion der Herzen wirbelloser Tiere. *Zool. Jahrb. Physiol.* **77**:477–668.

Roberts, A. (1968). Recurrent inhibition in the giant-fibre system of the crayfish and its effect on the excitability of the escape response. *J. Exp. Biol.* **48**:545–567.

Roberts, S.K. deF. (1960). Circadian activity rhythm in cockroaches. I. The freerunning rhythm in steady-state. *J. Cell. Comp. Physiol.* **55**:99–110.

Robertson, H.A. (1974). The innervation of the salivary gland of the moth, *Manduca sexta. Cell Tiss. Res.* **148**:237–245.

Robertson, H.A. (1975). The innervation of the salivary gland of the moth, *Manduca sexta:* Evidence that dopamine is the transmitter. *J. Exp. Biol.* **63**:413–419.

Robinson, G.R., and Scott, B.I.H. (1973). A new method of estimating micropipette tip diameter. *Experientia* **29**:1033–1034.

Roby, R.J., and Lettich, E. (1975). A simplified circuit for stimulus artifact suppression. *Electroencephal Ogr. Clin. Neurophysiol.* **39**:85–87.

Roeder, K.D. (1948). Organization of the ascending giant fiber system in the cockroach (*Periplaneta americana* L.) *J. Exp. Zool.* **108**:243–262.

Roeder, K.D., and Weiant, E.A. (1946). The site of action of DDT in the cockroach. *Science* **103**:304–306.

Roeder, K.D., and Weiant, E.A. (1948). The effect of DDT on sensory and motor structure of the cockroach leg. *J. Cell. Comp. Physiol.* **32**:175–186.

Rose, B. (1970). Junctional membrane permeability: Restoration by repolarizing current. *Science* **169**:607–609.

Roth. L.M. (1973). Inhibition of oöcyte development during pregnancy in the cockroach *Eublaberus posticus. J. Insect Physiol.* **19**:455–470.

Rowell, C.H.F. (1963a). The relation between stimulus parameters and current flow through stimulating electrodes. *J. Exp. Biol.* **40**:15–21.

Rowell, C.H.F. (1963b). A method for chronically implanting stimulating electrodes into the brains of locusts, and some results of stimulation. *J. Exp. Biol.* **40**:271–284.

Runion, H.I., and Pipa, R.L. (1970). Electrophysiological and endocrinological correlates during the metamorphic degeneration of a muscle fibre in *Galleria mellonella* (L.) (Lepidoptera). *J. Exp. Biol.* **53**:9–24.

Runion, H.I., and Usherwood, P.N.R. (1966). A new approach to neuromuscular analysis in the intact free-walking insect preparation. *J. Insect Physiol.* **12**:1255–1263.

Runion, H.I., and Usherwood, P.N.R. (1968). Tarsal receptors and leg reflexes in the locust and grasshopper. *J. Exp. Biol.* **49**:421–436.

Ryan, W.H., and Shankland, D.L. (1971). Synergistic action of cyclodiene insecticides with DDT on the membrane of giant axons of the American cockroach *Periplaneta americana* (L.). *Life Sci.* **10**:193–200.

Sakai, M. (1973). Nereistoxin on autonomic movement of the isolated hindgut of the American cockroach. *Appl. Entomol. Zool.* **8**:128–129.

Sandeman, D.C. (1968). A sensitive position measuring device for biological systems. *Comp. Biochem. Physiol.* **24**:635–638.

Sattelle, D.B. (1976). A simple assay for the synaptic actions of toxicological agents. In: *Evaluation of Biological Activity.* New York: Academic Press.

Sattelle, D.B., McClay, A.S., Dowson, R.J., and Callec, J.J. (1976). The pharmacology of an insect ganglion: Actions of carbamylcholine and acetylcholine. *J. Exp. Biol.* **64**:13–23.

Schaefers, G.A. (1966). The use of direct current for electronically recording aphid feeding and salivation. *Ann. Entomol. Soc. Am.* **59**:1022–1024.

Schechter, M.S., Dutky, S.R., and Sullivan, W.N. (1963). Recording circadian rhythms with a capacity-sensing device. *J. Econ. Entomol.* **56**:76–79.

Scheurer R., and Leuthold, R. (1969). Haemolymph proteins and water uptake in female *Leucophaea maderae* during the sexual cycle. *J. Insect Physiol.* **15**:1067–1078.

Schildmacher, H. (1950). Darmkanal und Verdauung bei Stechmückenlarven. *Biol. Zentralbl.* **69**:390–438.

Schneiderman, H.A., and Schechter, A.N. (1966). Discontinuous respiration in insects. V. Pressure and volume changes in the tracheal system of silkworm pupae. *J. Insect Physiol.* **12**:1143–1170.

Schofield, P.K., and Treherne, J.E. (1978). Kinetics of sodium and lithium movements across the blood–brain barrier of an insect. *J. Exp. Biol.* **74**:239–251.

Schroeder. W.J., Chambers, D.L., and Miyabara, R.Y. (1973). Mediterranean fruit fly: Propensity to flight of sterilized flies. *J. Econ. Entomol.* **66**:1261–1262.

Sevacherian, V. (1975). Activity and probing behavior of *Lygus hesperus* in the laboratory. *Ann. Entomol. Soc. Am.* **68**:557–558.

Shank, R.P., and Freeman, A.R. (1975). Cooperative interaction of glutamate and aspartate with receptors in the neuromuscular excitatory membrane in walking limbs of the lobster. *J. Neurobiol.* **6:**289–303.

Shankland, D.L. (1965). Nerves and muscles of the pregenital abdominal segments of the American cockroach, *Periplaneta americana* (L.). *J. Morphol.* **117:**353–386.

Shankland, D.L. (1976). The nervous system: Comparative physiology and pharmacology. In: *Insecticide Biochemistry and Physiology,* pp. 229–270. C.F. Wilkinson (Ed.). New York: Plenum.

Shankland, D.L., and Kearns, C.W. (1959). Characteristics of blood toxins in DDT-poisoned cockroaches. *Ann. Entomol. Soc. Am.* **52:**386–394.

Shankland, D.L., and Schroeder, M.E. (1973). Pharmacological evidence for a discrete neurotoxic action of dieldrin (HEOD) in the American cockroach, *Periplaneta americana* (L.). *Pestic. Biochem. Physiol.* **3:**77–86.

Shankland, D.L., Rose, J.A., and Donniger, C. (1973). The cholinergic nature of the cercal nerve-giant fiber synapse in the sixth abdominal ganglion of the American cockroach, *Periplaneta americana* L. *J. Neurobiol.* **2:**247–262.

Shipton, H.W., Emde, J.W., and Folk, G.E. (1959). A multiple point recorder for small animal locomotor activity. *Proc. Iowa Acad. Sci.* **66:**407–412.

Singh, K.M., Pradhan, S., and Dakshinamurthy, C. (1972). Selection of suitable nerve preparations of *Schistocerca gregaria* Forsk. for electrophysiological studies with performance of difference locust salines. *Ind. J. Entomol.* **34:**155–168.

Sláma, K. (1976). Insect hemolymph pressure and its determination. *Acta Entomol. Bohemoslov.* **73:**65–75.

Smith, E.N., and Campbell, G.D. (1975). Direction and size discriminating activity recorder. *Environ. Entomol.* **4:**980–982.

Smith, J.J.B., and Friend, W.G. (1970). Feeding in *Rhodnius prolixus:* Responses to artificial diets as revealed by changes in electrical resistance. *J. Insect Physiol.* **16:**1709–1720.

Smyth, T., Greer, M.N., and Griffiths, D.J.G. (1973). Insect neuromuscular synapses. *Am. Zool.* **13:**315–319.

Snodgrass, R.E. (1956). *The Anatomy of the Honeybee.* Ithaca: Comstock.

Sokolove, P.G. (1975). Locomotory and stridulatory circadian rhythms in the cricket, *Teleogryllus commodus. J. Insect Physiol.* **21:**537–558.

Spangler, H.G. (1969). Suppression of honeybee flight activity with substrate vibration. *J. Econ. Entomol.* **62:**1185–1186.

Spencer, H.J. (1971a). Programmable nanoampere constant current sources for iontophoresis. *Med. Biol. Eng.* **9:**683–702.

Spencer, H.J. (1971b). A linearly reading micro-electrode resistance meter. *Electroencephalogr. Clin. Neurophysiol.* **31:**518–519.

Spencer, H.J. (1972). An epochal ratemeter for neurophysiological studies. *Electroencephalogr. Clin. Neurophysiol.* **33:**228–231.

Spencer, H.J. (1973). A microlathe for constructing miniature multibarrel micropipettes for iontophoretic drug application. *Experientia* **29:**1577–1579.

Spencer, H.J. (1975). Micro iontophoresis techniques. *Carrier* **2:**1–7. (Published by David Kopf Instruments.)

Spira, M.E., Yarom, Y., and Parnas, I. (1976). Modulation of spike frequency by regions of special axonal geometry and by synaptic inputs. *J. Neurophysiol.* **39:**882–899.

S.-Rózsa, K., and Véró, M. (1971). Electrocardiograms in insecta and gastropoda and their changes under experimental conditions. *Annal. Biol. Tihany* **38**:79–86.

Stange, G., and Hardeland, R. (1970). Eine Methode zur Registrierung dur Laufaktivität von kleinen Insecten. *Oecologia* **5**:400–405.

Starratt, A.N., and Brown, B.E. (1975). Structure of the pentapeptide proctolin, a proposed neurotransmitter in insects. *Life Sci.* **17**:1253–1256.

Steel, C.G.H. (1978). Some functions of identified neurosecretory cells in the brain of the aphid, *Megoura viciae*. *Gen. Comp. Endocrinol.* **34**:219–228.

Stout, J.F. (1971). A technique for recording the activity of single interneurons from freely-moving crickets (*Gryllus campestris* L.). *Z. Vgl. Physiol.* **74**:26–31.

Sykes, J.M., Ives, A.G., and Rothwell, G.P. (1970). An apparatus for recording internal stresses in electrodeposits during plating. *J. Phys.* **3**:941.

Szymanski, J.S. (1914). Eine Methode zur Untersuchung der Ruhe-und Aktivitätsperioden bei Tieren. *Pfluegers Arch. Gesamte Physiol. Menschen Tiere* **158**:343–385.

Tachibana, K., and Nagashima, C. (1957). Studies on the heartbeat of Insect. *Nippon ogo dobutsu konschugakkai shi* **1**:155–163 (text Japanese, summary English).

Takahashi, S. (1934). Studies on the action current of the dorsal vessel of the silkworm. *Physiol. Papers. Tokyo Jikeikai Med. Coll.* **3**:387.

Tasaki, K., Tsukahara, Y., and Ito, S. (1968). A simple, direct and rapid method for filling microelectrodes. *Physiol. Behav.* **3**:1009–1010.

Ten Cate, J. (1924). Contribution á la physiologie comparée du tube digestif. III. Les movements rhythmiques spontanés de l'oesophage isolé et du gosier de *Dytiscus marginalis*. *Arch. Neerl. Physiol.* **9**:598–604.

Tenney, S.M. (1953). Observations on the physiology of the Lepidopteran heart with special reference to the reversal of beat. *Oecologia* **3**:286–306.

Thomas, A., and Mesnier, M. (1973). Le rôle du systéme nerveux central sur les mécanismes de l'oviposition chez *Carausius morosus* et *Clitumnus extradentatus*. *J. Insect Physiol.* **19**:383–396.

Thomas, M.V. (1976). Insect blood–brain barier: An electrophysiological investigation of its permeability to the aliphatic alcohols. *J. Exp. Biol.* **64**:101–118.

Thomas, M.V., and Treherne, J.E. (1975). An electrophysiological analysis of extra-axonal sodium and potassium concentration in the central nervous system of the cockroach (*Periplaneta americana* L.). *J. Exp. Biol.* **63**:801–811.

Thomson, A.J. (1975). Synchronization of function in the foregut of the blowfly *Phormia regina* (Diptera: Calliphoridae) during the crop-emptying process. *Can. Entomol.* **107**:1193–1198.

Ting, K.Y., and Brooks, M.A. (1965). Sodium potassium ratios in insect cell culture and the growth of cockroach cells. *Ann. Entomol. Soc. Am.* **58**:197–202.

Treherne, J.E. (1961). Sodium and potassium fluxes in the abdominal nerve cord of the cockroach, *Periplaneta americana*. *J. Exp. Biol.* **38**:315–322.

Treherne, J.E., and Foster, W.A. (1977). Diel activity of an intertidal beetle, *Dicheirotrichus gustavi* Crotch. *J. Anim. Ecol.* **46**:127–138.

Treherne, J.E., and Willmer, P.G. (1975). Hormonal control of integumentary water loss: Evidence for a novel neuroendocrine system in an insect (*Periplaneta americana*). *J. Exp. Biol.* **63**:143–159.

Unwin, D.M. (1978). A versatile high frequency radio microcautery. *Physiol. Entomol.* **3**:71–73.

Uramoto, S. (1932). On the mechanogram and electrogram of the dorsal vessel and the alimentary canal in the silkworm larva. *Bull. Imp. Seric. Exp. Sta. (Tokyo)* **8**:121–134 (in Japanese; English summary).

Usherwood, P.N.R. (1967). Insect neuromuscular mechanisms. *Am. Zool.* **7**:553–582.

Usherwood, P.N.R. (1968). A critical study of the evidence for peripheral inhibitory axons in insects. *J. Exp. Biol.* **49**:201–222.

Usherwood, P.N.R. (1969). Electrochemistry of insect muscle. *Adv. Insect Physiol.* **6**:205–278.

Usherwood, P.N.R. (1972). Transmitter release from insect excitatory motor nerve terminals. *J. Physiol.* **227**:527–551.

Usherwood, P.N.R. (1973). Action of iontophoretically applied gamma aminobutyric acid on locust muscle fibres. *Comp. Biochem. Physiol.* **44A**:663–664.

Usherwood, P.N.R. (1974). Nerve–muscle transmission. In: *Insect Neurobiology*, pp. 245–305. J.E. Treherne (Ed.). Amsterdam: North-Holland.

Usherwood, P.N.R. (1975). *Insect Muscle*. London: Academic Press.

Usherwood, P.N.R., and Grundfest, H. (1965). Peripheral inhibition in skeletal muscle of insects. *J. Neurophysiol.* **28**:497–518.

Usherwood, P.N.R., and Machili, P. (1968). Pharmacological properties of excitatory neuromuscular synapses in the locust. *J. Exp. Biol.* **49**:341–361.

Usherwood, P.N.R., and Runion, H.I. (1970). Analysis of the mechanical responses of metathoracic extensor tibiae muscle of free-walking locusts. *J. Exp. Biol.* **52**:39–48.

Usherwood, P.N.R., Runion, H.I., and Campbell, J. (1968). Structure and physiology of a chordotonal organ in the locust leg. *J. Exp. Biol.* **48**:305–323.

Uvarov. B.P. (1966). *Grasshoppers and Locusts*. A handbook of general acridology in two volumes. Cambridge: Cambridge University Press.

Vedel, F., Bernard, U., Boussinesq, J., Morin, J., Nazet, C., and Patou, C. (1971). Disjoncteurs par explosifs. Commutation ultra-rapide de courants intenses supérieurs au méga-ampére. *Rev. Gen. Elec.* **80**:873–877.

Vernon, J.E.N. (1973). Physiological control of the chromatophores of *Austrolestes annulosus* (Odonata). *J. Insect Physiol.* **19**:1689–1703.

Véró, M. (1971). Negative capacitance amplifier for microelectrode investigations. *Annal. Biol. Tihany* **38**:107–115.

Véró, M. (1972). Transistorized square wave generator for biological investigations. *Annal. Biol. Tihany* **39**:75–80.

Véró, M., and Salanki, J. (1969). Inductive attenuator for continuous registration of rhythmic and periodic activity of mussels in their natural environment. *Med. Biol. Eng.* **7**:235–237.

Visser, B.J., Spanjer, W., deKlonia, H., Piek, T., van der Meer, C., and van der Drift, A.C.M. (1976). Isolation and some biochemical properties of a paralysing toxin from the venom of the wasp *Microbracon hebetor* (Say). *Toxicon* **14**:357–370.

Vowles, D.M. (1964). Models in the insect brain. In: *Neural Theory and Modeling*, R.F. Reiss (Ed.). Stanford, California: Stanford University Press.

Wallengren, H. (1914). Physiologisch–Biologische Studien über die Atmung bei den Arthropoden. II. Die Mechanik der Atembewegungen bei *Aeschna* larven. *Acta Univ. Lund.* **10:**1–24.

Walther, W., and Rathmayer, W. (1974). The effect of *Habrobracon* venom on excitatory neuromuscular transmission in insects. *J. Comp. Physiol.* **89:**23–38.

Wang, C.M., Narahashi, T., and Yamada, M. (1971). The neurotoxic action of dieldrin and its derivatives in the cockroach. *Pestic. Biochem. Physiol.* **1:**84–91.

Wann, K.T., and Goldsmith, M.W. (1972). The pH dependence of the micro-electrode tip potential. *J. Physiol* **227:**57P.

Wareham, A.C., Duncan, C.J., and Bowler, K. (1973). Bicarbonate ions and the resting potential of cockroach muscle: Implications for the development of suitable saline media. *Comp. Biochem. Physiol.* **45A:**239–246.

Wareham, A.C., Duncan, C.J., and Bowler, K. (1974a). The resting potential of cockroach muscle membrane. *Comp. Biochem. Physiol.* **48A:**765–797.

Wareham, A.C., Duncan, C.J., and Bowler, K. (1974b). Electrogenesis in cockroach muscle. *Comp. Biochem. Physiol.* **48A:**799–813.

Wasserthal, L.T. (1975). Herzschlag-Umkehr bei Insekten und die Entwicklung der imaginalen Herzrhythmik. *Verh. Dtsch. Zool. Ges.* **1974:**95–99.

Wasserthal, L.T. (1976). Heartbeat reversal and its coordination with accessory pulsatile organs and abdominal movements in Lepidoptera. *Experientia* **32:**577–578.

Weiss, M.D. (1973). *Biomedical Instrumentation.* Philadelphia: Chilton Book Co.

Weis-Fogh, T. (1956). Biology and physics of locust flight. II. Flight performance of the desert locust (*Schistocerca gregaria*). *Philos. Trans. R. Soc. London Ser. B* **239:**459–451.

West, L.S. (1951). *The Housefly.* Ithaca: Comstock Publ. Co.

West, L.S., and Peters, O.B. (1972). *Annotated Bibliography of* Musca domestica *Linnaeus.* London: Dawsons of Pall Mall.

Wheal, A.V., and Kerkut, G.A. (1975). The antagonistic action of L-glutamate diethyl ester on the neuromuscular junction. *Comp. Biochem. Physiol.* **51C:**79–81.

Wheeler, W.M. (1910). *Ants.* New York: Columbia University Press.

Whitehead, A.T. (1971). The innervation of the salivary gland in the American cockroach: Light and electron microscopic observations. *J. Morphol.* **135:**483–506.

Whitehead, A.T. (1973). Innervation of the American cockroach salivary gland: Neurophysiological and pharmacological investigations. *J. Insect Physiol.* **19:**1961–1970.

Wigglesworth, V.B. (1953). *The Principles of Insect Physiology,* fifth ed. London: Methuen.

Wilkens, L.A., and Wolfe, G.E. (1974). A new electrode design for *en passant* recording, stimulating, and intracellular dye infusion. *Comp. Biochem. Physiol.* **48A:**217–220.

Wille, J. (1920). Biologie und Bekämpfung der deutschen Schabe (*Phyllodromia germanica* L.). *Z. Angew Entomol.* **7:**1–140.

Willey, R.B. (1961). The morphology of the stomadeal nervous system in *Periplaneta americana* (L.) and other Blattaria. *J. Morphol.* **108**:219–261.

Williams, C.M., and Chadwick, L.E. (1943). Technique for stroboscopic studies of insect flight. *Science* **98**:522–524.

Williams, G.C., Ballard, R.C., and Hall, S. (1968). Mechanical movement of the insect heart recorded with a continuous laser beam. *Nature (London)* **220**:1241–1242.

Wills, W., Carroll, D.F., and Jones, G.E. (1974). A simple method for artificially feeding mosquitoes. *Mosquito News* **34**:119–121.

Wilson, D.M. (1968). The central nervous control of insect flight and related behaviour. *Adv. Insect Physiol.* **5**:289–338.

Wise, R.M. (1975). Capacitive transducer senses tension in muscle fibers. *Electronics* (June 26, 1975), **48**:(13), 97.

Wong, R.K.S., and Pearson, K.G. (1976). Properties of the tochanteral hair plate and its function in the control of walking in the cockroach. *J. Exp. Biol.* **64**:233–249.

Yamasaki, T., and Narahashi, T. (1959). The effects of K$^+$ and Na$^+$ ions on the resting and action potentials of the cockroach giant axon. *J. Insect Physiol.* **3**:146–158.

Yeager, J.F. (1931). Observation on crop and gizzard movements in the cockroach, *Periplaneta fuliginosa. Ann. Entomol. Soc. Am.* **24**:739–745.

Yeager, J.F. (1938). Mechanographic method of recording insect cardiac activity, with reference to effect of nicotine on isolated heart preparations of *Periplaneta americana* Linnaeus. *J. Agric. Res.* **56**:267–276.

Yeager, J.F. (1939). Electrical stimulation of isolated heart preparations from *Periplaneta americana* Linnaeus. *J. Agric. Res.* **59**:121–132.

Yeager, J.F., and Gahan, J.B. (1937). Effects of the alkaloid nicotine on the rhythmicity of isolated heart preparations from *Periplaneta americana* and *Prodenia eridania. J. Agr. Res.* **55**:1–19.

Yeager, J.F., and Hager, A. (1934). On the rates of contraction of the isolated heart and Malpighian tube of the insect, *Periplaneta orientalis.* method. *Iowa State Coll. J. Sci.* **8**:391–395.

Yeager, J.F., and Swain, R.B. (1934). An entomotograph, an instrument for recording the appendicular or locomotor activity of insects. *Iowa State Coll. J. Sci.* **8**:519–525.

Young, D. (1973). Specificity and regeneration in insect motor neurons. In: *Developmental Neurobiology of Arthropods.* D. Young (Ed.). Cambridge: University Press.

Young, S. (1973). *Electronics in the Life Sciences.* New York: Wiley.

Zimmerman, W.F., Pittendrigh, C.S., and Pavlidis, T. (1968). Temperature compensation of the circadian oscillation in *Drosophila pseudoobscura* and its entrainment by temperature cycles. *J. Insect Physiol.* **14**:669–684.

Appendix

Ace Glass Co., 1430 NW Boulevard, Vineland, New Jersey 08360

Acme Slotted Angle, Interlake Steel Corp., 135th Street and Perry Avenue, Chicago, Illinois 60627

Alfa Organics Ventron 8 Congress Street, Beverly, Massachusetts 01915

Alfa Wire Corp., Alfa Machine and Tool Co., Inc., 19 Just Road, Fairfield, New Jersey 07006

Akers Electronic Co., Horten, Norway

Allied Control Company, Inc., 100 Relay Road, Plantsville, Connecticut 06479

Allco, Société D'instrumentation électronique, Allard et Compagnie, 57, Rue Saint Sauveur, 91160 Ballainvilliers B.P. 31 - Longjumeau 91, France

Amalgamated Dental Co., Distributors Ltd., 26-40 Broadwink Street, London W1A 2AD, England

American Optical Corp., Fiber Optics Division, Southbridge, Massachusetts 01550

American Optical Corp., Scientific Instrument Division, Buffalo, New York 14215

Annex Research, P.O. Box 15044, Santa Ana, California 92705

Ampex Corp., Instrumentation Division, 401 Broadway, Redwood City, California 94063

Applied Fiberoptics and Scientific Specialities, Inc., 46 River Street, Southbridge, Massachusetts 01550

Ardel Kinematic, 125-20 18th Avenue, College Point, New York 11356

Argonaut, 3106 S.W. 87th Avenue, Portland, Oregon 97225

Arnold R. Horwell, Ltd., Laboratory & Clinical Supplies, 2 Grangeway Kilburn High Road, London NW6 lYB, England

Ash, see Amalgamated Dental Co.

AUS Jena, Jenoptik Jena G.m.b.H., DDR 69 Jena, Karl Zeiss Strasse 1, E. Germany (dist. by International Micro-Optics)

Bailey Instruments Co., 515 Victor Street, Saddlebrook, New Jersey 07662

Barr & Stroud, Annieslandcross, Glasgow, Scotland

Bausch & Lomb, Scientific Optical Products Division, 38804 Bausch Street, Rochester, New York, 14602

BDH Chemical Ltd., see British Drug House

Beckman Instruments Co., 2500 Harbor Boulevard, Fullerton, California 92634

Belden, 2000 S. Batavia Avenue, Geneva, Illinois 60134

Bell and Howell, 360 Sierra Madre Villa, Pasadena, California 91109

Berk-Tek, Inc., P.O. Box 60, R.D. 1, Reading, Pennsylvania 19607

BIOCOM Inc., 9522 West Jefferson Boulevard, Culver City, California 90230

Bioelectric Instruments, Inc., Hastings-on-Hudson, New York 10706

Biolab Ltd., 2-10 Regent Street, Cambridge CD2 1DB, England

Biomation, 10411 Bub Road, Cupertino, California 95014

Biomedical Electronics, Inc., 653 Lofstrand Lane, Rockville, Maryland 20850

B & K Instruments, Inc., 5111 West 164th Street, Cleveland, Ohio 44142

BLH Electronics Inc., 42 Fourth Avenue, Waltham, Massachusetts 02154

Borden Co., Chemical Division, 180 E. Broad, Columbus, Ohio 43215

Brinkmann Instruments, Cantiague Road, Westbury, New York 11590

British Drug House, Poole, Dorset

BRS/LVE, 5301 Holland Drive, Beltsville, Maryland 20705

Brush Instruments, see Gould

Buckbee-Mears Co., 1900 American National Bank Bldg., St. Paul, Minnesota 55101

Burkhard Mfg. Co., Ltd., Rickmansworth, Hertfordshire WD3 lPJ, England

Burr-Brown Research Corp., International Airport Industrial Park, Tucson, Arizona 85706

Caltron Industries, 2015 Second Street, Berkeley, California 94710

Carl Zeiss, Inc., 444 Fifth Avenue, New York, New York 10018

Carl Zeiss, Ltd., 31-36 Foley, London WlP 8AP, England

CENCO, Central Scientific Company of California, 6446 Telegraph Rd., Los Angeles, California 90022

C.F. Palmer, Ltd., Lane End Road, High Wycombe, Bucks., England

Chemical Rubber Co., Chemprene, Division of the Richardson Company, 570 Fishkill Avenue, Beacon, New York 12508

Circon Corporation, 749 Ward Drive, Santa Barbara, California 93111

Clair Armin, Karlsbad Insect Pins, 191 W. Palm Avenue, Reedley, California 93654

Clark Electromedical Instruments, P.O. Box 8, Pangbourne, Reading RG8 7HU, England

Clarke Biomedical Instruments, (see Clark Electromedical Instruments)

Clay Adams, Div. of Becton, Dickinson & Co. (B-D), Parsippany, New Jersey 07054

Cole-Parmer Instrument Co., 7425 N. Oak Park Avenue, Chicago, Illinois 60648

Collins Electric, Old Love Point Road, Stevensville, Maryland 21666

Columbus Instruments, P.O. Box 5244, Columbus, Ohio 43212

Comark Electronics, Brookside Ave., Rustington, Sussex BN16 3LF, England

Consolidated Reactive, CRM, A Division of Consolidated Refining Co., Inc., 115 Hoyt Avenue, Mamaroneck, New York 10543

Cooner Sales Co., 9129 Lurline Avenue, Chatsworth, California 91311

Corning Glass Works, Corning, New York 14830

Coronet Imp, Cleveland Road, Dalton, Georgia 30720

Curtin Scientific, 4220 Jefferson, Houston, Texas 77023

C.W. Dickey Associates, P.O. Box 71, State College, Pennsylvania 16801

C.Z. Scientific Instruments Ltd., 2, Elstree Way, Boreham Wood, Hertz WD6 lNH, England

Dagan, Inc., 201 Minehaha Avenue, Minneapolis, Minnesota 55404

Dallas Instruments, Inc., P.O. Box 38189, Dallas, Texas 75238

Data Inc., 732 South Federal Street, Chicago, Illinois 60605

Data Inc., P.O. Box 2071, 130 South Mason, Fort Collins, Colorado 80521

David Kopf Instruments, 7324 Elmo Street, Tujunga, California 91042

Daytronic Corporation, 2875 Culver Avenue, Dayton Ohio 45429

Devcon, Ltd., Theale, Berks, England

Devcon Ltd., Danvers, Massachusetts 01923

Devcon Canada Ltd., Scarborough, Ontario, Canada

Dexter Corp., Hysol Division, Olean, New York 14760

Devices Ltd., 13-15 Broadwater Road, Welwyn Garden City, Hertz, England AL7 3AP

Digital Equipment Corp., 146 Main Street, Maynard, Massachusetts 01754

Digitimer Ltd., Medical Systems Corp., 230 Middle Neck Road, Great Neck, New York, 11021; 37 Hydeway, Welwyngarden City, Hertz, AL7 3BE, England

Dircon Corp. Santa Barbara Airport, Goleta, California 93017

Dolan-Jenner Industries, Inc., 200 Ingalls Court, Melrose, Massachusetts, 02176

D-MAC Ltd., Shaftesbury Street, High Wycombe, HP11 2NA Bucks, England

Dorsch Elektronik, München 8, Worthstr. 8, Germany

Dow Corning Corp., Electronic Materials Dept., Midland, Michigan 48640

Dow Corning Silicone Adhesives, Dow Corning Corp, Bonding, Sealing, Adhering Products, P.O. Box 1592, Midland, Michigan 48640

Down Bros. and Mayer & Phelps Ltd., Church Path, Mitcham Surrey CR4 3UE, England

Down Bros. and Mayer & Phelps Ltd., 261 Davenport Road, Toronto 180, Ontario, Canada

Downs Surgical, Inc., 655 - 74th Street, Niagara Falls, New York 14304

Driver-Harris, 201 Middlesex Street, Harrison, New Jersey 07029

Drummond Scientific Co., 500 Parkway, Broomall, Pennsylvania 19008

Dumont Oscilloscope Laboratories, Inc., 40 Fairfield Place, West Caldwell, New Jersey, 07006

Ealing Corporation, 22 Pleasant Street, South Natick, Massachusetts 01760 (agent for C.F. Palmer Ltd.)

EDCO Scientific, Inc., 65 Tosca Drive, Stoughton, Massachusetts 02072

ED&G, Environment Equip. Div., 151 Bear Hill Road, Waltham, Massachusetts 02154

Eico, 283 Malta Street, Brooklyn, New York 11207

Electrisola (see Elektroisola)

Elektroisola, Dr. G. Schildbach, D 5381 Eckenhagen, West Germany.

Electronics for Life Sciences, Inc., P.O. Box 697, Rockville, Maryland 20851

Engelhard Industries Division, 430 Mountain Avenue, Murray Hill, New Jersey 07974

Endevco, Rancho Viejo Road, San Juan Capistrano, California 92675

Ernest Fullam, Inc., P.O. Box 444, Schenectady, New York 12301

Ernst Leitz, Inc. (GMBH), D-6330 Wetzlar, Germany

Eric Sobotka, 110 Finn Court, Farmingdale, New York 11735

Ethicon, Ltd., P.O. Box 408, Bankhead Avenue, Edinburgh EH11 4HE Scotland

Exact Electronics, Inc., Box 160, Hillsboro, Oregon 97123

E.W. Wright, 760 Durham Road, Guilford, Connecticut 06437

Farnell Instruments Ltd., Sanbeck Way Wetherby, Yorkshire LS22 4DH, England

Faulhaber, Monroeville, Ohio 44847

Fenwall Electronics, Inc., 63 Fountain Street, Framingham, Massachusetts, 01701

Fiberglass Industries, Inc., Amsterdam, New York 12010

Fisher Scientific Co. (overseas branch), Zeltweg 67, 8032 Zurich, Switzerland

Fisher Scientific Co., 711 Forbes Avenue, Pittsburgh, Pennsylvania 15219

Flexi-Optics Laboratories, Inc., 269 Mechanic Street, Marlboro, Massachusetts 01752

Flex-set, see Wilks Instruments

Firma Osram, 1 Berlin N 65, Groninger Strasse 19

Frederick Haer & Co., PO. Box 2138, Ann Arbor, Michigan 48106

Frederick Haer & Co., P.O. Box Cheltenham Glos. GL519SA

Freiedrick & Dimmock, Box 230, Millville, New Jersey 08332

Furniglas Ltd., 136-138 Great North Road, Hatfield, Hertz, England

GAF-APSCO of Illinois Inc., P.O. Box 192, Woodstock, Illinois 60098

Gallenkamp/ Schiff, 35 N. 4th, Columbus, Ohio 43215

Garner Glass Co., 1770 South Indian Hill, Claremont, California 91711

G. C. Electronics, 400 South Wyman Street, Rockford, Illinois 61101

General Electric Co., Cleveland Wire Plant, P.O. Box 3050, Cleveland, Ohio 44117

General Electric Co., 17000 E. Gale Ave., City of Industry, California 91745

General Electric, Sheet Products Section, Chemical and Metallurgical Division, 1 Plastics Avenue, Pittsfield, Massachusetts 01201

General Electric Silicone Adhesives Products Dept., Waterford-Mechanicsville Road, Waterford, New York 12118

G. H. Bloore, Ltd., 480 Honeypot Lane, Stanmore, Middlesex HA7 1Jt, England

Gibson Medical Electronics, Inc., P.O. Box 27, Middleton, Wisconsin 53562

GLASFABRIK, Postfach 9, Malsfeld 3509, Germany

Glass Company of America, Inc., Oakland and Ridge Avenues, Bargintown, New Jersey 08232

Goodfellow Metals Ltd., Ruxley Towers, Claygate-Esher, Surrey, KT10 OTS, England

Gould, Inc., 3631 Perkins Avenue, Cleveland, Ohio 44114

Grass Instrument Co., 101 Old Colony Avenue, Quincy, Massachusetts 02169

Gulton Industries, Recorder Systems Div., Gulton Industrial Park, East Greenwich, Rhode Island 02818

Hacker Instruments Inc., P.O. Box 646, Fairfield, New Jersey 07006

Hamilton Bell Co., Montvale, New Jersey 07645

Harbutt's Plasticine, Ltd., Bath, England

Harry's Radio, 303 Edgeware Road, London W2 1BW, England

Harvard Apparatus Co., Inc., 150 Dover Road, Millis, Massachusetts 02054

Heat-Prober Division, William Wahl Corp., 12908 Paname Street, Los Angeles, California 90066

Heat Systems-Ultrasonics, Inc., 38 E. Mall, Plainview, New York 11803

Heath/Schlumberger Instruments, Benton Harbor, Maryland 49022

Henry's Radio, Ltd., 303 Edgeware Road, London W2 IBN, England

Herald Electronics, Lincolnwood, Illinois 60645

Hewlett Packard, 1217 Megrin-Geneva, Switzerland

Hewlett Packard, 1501 Page Mill Road, Palo Alto, California 94304

Hilgenberg, K. (see Glasfabrik)

Hocking Anchor Hocking Co., 109 N. Broad, Lancaster, Ohio 43130

Hoffman, P.R. Co. (Lapidary) 350 Cherry St., Carlisle, Pennsylvania 17013

Harvard Apparatus Co., Inc., 150 Dover Road, Millis, Massachusetts 02054

Industrial Science Associates, Inc., 63-15 Forest Avenue, Ridgewood, New York 11227

Insl-x Co., Ossining, New York 10562

International Micro-Optics, Div. of Charvoz-Carsen Corp., 5 Daniel Road, Fairfield, New Jersey, 07006 (dist. for AUS Jena)

in Vivo Metric System, P.O. Box 217, Redwood City, California 95470

Isleworth Electronics, Frederick Street., Waddesdon, Bucks, England

Jenoptik (see AUS Jena)

J.L. Hammett Co., Hammett Place, Braintree, Massachusetts 02184

Johnson & Matthey Ltd., 73-83 Hatton Garden, London, F.E.C.I. England

John Weiss & Son, Ltd., see Weiss

Keystone Optical Fibers, 151 Hallet St., Boston, Massachusetts 02124

Krell Electronics, 64 Sylvan Avenue, Clifton, New Jersey 07011

Kressilk Products, Inc. 420 Sawmill River Road, Elmsford, New York 10523

Labtron Scientific Corp., 341 Conklin St., Farmingdale, New York 11735

Ladd and Fullam, Ladd Research Industries, Inc. Burlington, Vermont 05401

McGaw Laboratories, Div. of American Hospital Supply Corp., 1015 Grandview Avenue, Glendale, California 91201

Med Associates, Box 1506, Allentown, Pennsylvania 18105

Medical Electronics & Equipment News, Chilton Way, Radnor, Pennsylvania 19089

Medwire Corp., 121 South Columbus Avenue, Mount Vernon, New York 10553

Mentor Corp., 3104 West Lake Street, Minneapolis, Minnesota 55416

Micromanipulator Co., 1120 S.W. Industrial Avenue, Escondido, California 92025

Molecule Wire Corp (The), P.O. Box 495, Farmingdale, New Jersey 07727

Mupac Corp., 646 Summer Street, Brockton, Massachusetts, 02402

Narishige Scientific Instruments (see Labtron), 1754-6 Karasuyama-Cho Setagaya-Ku, Tokyo, Japan

Neurolog, see Digitimer

Newark, 4747 W. Century Blvd., Inglewood, California 90304

New Metals and Chemicals, Ltd., Chaucery House, Chaucery Lane, London, W.C. 2, England

Niagra Electron Laboratories, 1-11 Rochambeau Avenue, Andover, New York 14806

Nicolet Instrument Corp., 5225 Verona Road, Madison, Wisconsin 53711

Nicolet Instruments Ltd., 80A Emscote Road, Warwick, England

Nicolet Instrument G.m.b.H., Goerdeler Strasse 48, D-605 Offenbach am Main, West Germany

Nihon Waters Ltd., 3-4-4 Lidabashi, Chiyoda-Ku, Tokyo, Japan

Nikon Instrument Group, Ehrenreich Photo-Optical Industries, Inc., 623 Stewart Avenue, Garden City, New York 11530

Northern Scientific, Inc., 2551 West Beltline, P.O. Box 66, Middleton, Wisconsin 53562

Ohmite Manufacturing Co., 3601 Howard Street, Skokie, Illinois 60076

Olympus Corporation of America Precision Instrument Division, 2 Nevada Drive, New Hyde Park, New York 11040

Omega Engineering, Inc., Box 4047, Stamford, Connecticut 06907

Ortec Inc., Life Science Products, 100 Midland Road, Oak Ridge, Tennessee 37830

Pacific Stereo Stores, Herald Electronics, Lincolnwood, Illinois 60645

Palmer (See C. F. Palmer and Ealing)

Parrys (Tools) Ltd., 325-329 Old Street, London E.C.1., England

PBL International, Inc., P.O. Box 108, Newburyport, Massachusetts 01950

Pelco (Ted Pella Co.)., P.O. Box 510, Tustin, California 92680

Pemtek, Inc., 942 Commercial Street, Palo Alto, California 94303

Philips Electronic Instruments (A sub. of North American Philips Corp.) 750 S. Fulton Avenue, Mt. Vernon, New York 10550

Philips Test & Measuring Instruments, Inc., 224 Duffy Avenue, Hicksville, Long Island, New York 11802

Phipps & Bird, Inc., 8741 Landmark Road, Richmond, Virginia 23261

Physical Data, Inc. 5160 North Lagoon, Portland, Oregon 97217

Polysciences, Inc., Paul Valley Industrial Park, Warrington, Pennsylvania 18976

Popper & Sons, Inc., 300 Denton Avenue, New Hyde Park, New York 11040

W.R. Prior & Co., Ltd., London Road, Bishop's Stortford, Herts., England (see also Eric Sobotka Co.)

Pye Dynamics Ltd., Park Avenue, Bushey, Herts., England

Radio Shack, 2617 W. Seventh, Fort Worth, Texas 76107

Radio Spares P.O. Box 427 13-17 Epworth Street, London EC2P 2HA, England

Rank-Kershaw Fibrox, The Rank Organisation Ltd., A. Kershaw and Sons, Millbank Tower, Millbank, London SW1

Rank Optics, 200 Harehills Lane, Leeds LS8 5QS, England

Rank Precision Industries Ltd., Industrial Division (Rank-Kershaw) 200 Harehills Lane, Leeds LS8 5QS, England

Reactor Experiments, Inc., 963 Terminal Way, San Carlos, California 94070

Reichardt Bayerische Hypotheken-und Wechsel-Bonk D-8000 München 2, Postfach 20 05 27

Reynolds Aluminum, 5670 Wilshire Boulevard, Los Angeles, California 90036

R. I. Ltd. Rescarch Instruments, Ltd. Kernick Road, Penryn, Cornwall, England

Roboz, 155 Club Road, Suite A, Pasadena, California 91105

Rocol House, Swillington, Leeds LS28 8BS, England

Rocol R.T.D., see William & Nye, Inc., see also Rocol House

Rohm & Haas Co., Plastics Dept., Independence Mall-West, Philadelphia, Pennsylvania 19105

Sangamo Electric Co., P.O. Box 3347, Springfield, Illinois 62708

Schaevitz Engineering Inc., P.O. Box 505, Camden, New Jersey 08101

Schott Optical Glass Inc. 400 York Avenue, Duryea, Pennsylvania 18642

Scientific Products (Division of American Hosptial Supply), 1430 Waukegan Road, McGaw Park, Illinois 60085

Scientific & Research Instruments Ltd., 335 Whitehorse Road, Croydon, Surrey, England

Scientific & Research Instruments Ltd., Fircroft Way, Edenbridge, Kent TN8 6HE, England

Secon Metals Corp., Secon Wire Division, 5-7 Intervale St., White Plains, New York 10606

Sensorex, 17502 Armstrong Avenue, Irvins, California 92705

Shockman Instruments, Ltd. Lane End Road, High Wycombe, Bucks, England

Shockman Instruments, Ltd., Mineral Lane, Chesham, Bucks., HP5 1MU, England

Siemens Corp., 186 Wood Avenue South, Iselin, New Jersey 08830

Sigmund Cohn (Medwire), 125 S. Columbus Avenue, Mount Vernon, New York 10553

Silicone Products (GE), Waterford, New Jersey 12188

Skan-A-Matic Corporation, P.O. Box S, Route 5, Elbridge, New York 13060

Small Parts, Inc., 6901 N.E. Third Avenue, Miami, Florida 33138

Smiths Ltd., Lewcos House, Lwr. Wortley Road, Leeds, England

Société D'instrumentation électronique, (see Allco)

Spare Parts, Inc., Blackburn, Lancs., England

SRI Scientific Resources, Inc. 3300 Commercial Avenue, Northbrook, Illinois 60062

Standard Wire and Cable, 3440 Overland Avenue, Los Angeles, California 90034

Stoelting Co., 1350 S. Kostner Avenue, Chicago, Illinois 60623

Sutter Instrument Co., P.O. Box 16385, San Francisco, California 94116

Sylvania Electric Products, Inc., Chemical and Metallurgical Division, Towanda, Pennsylvania 18348

Strickland Electric Co., 1427 18th Avenue, Columbus, Ohio 43211

Takeda Riken Industry Co. Ltd., 1-32-1 Asahi-cho, Nerima-ku, Tokyo 176, Japan

Tandberg of America, Inc, 8 Third Avenue, Pelham, New York 10803

TEAC Corporation of America, Technical Products, 7733 Telegraph Road, Montebello, California 90640

Tektronix, Inc., P.O. Box 500, Beaverton, Oregon 97005

Terrasyn Inc., P.O. Box 975, Longmont, Colorado 80501

Tescom Corp., 2600 Niagara Lane N, Minneapolis, Minnesota 55441

Test & Measuring Instruments Inc., 224 Duffy Avenue, Hicksville, New York 11802

Texas Instruments Inc., Digital Systems Division, P.O. Box 1444, Huston, Texas 77001

Thomas Apparatus (Arthur H. Thomas Co.), Vine Street at Third, P.O. Box 779, Philadelphia, Pennsylvania 19105

3M Co., Data Recording Products Division, 3-M Center Building, 224-61, St. Paul, Minnesota 55101

3M Co., Medical Products Division, 3-M Center Building, 220-7W, St. Paul, Minnesota 55101

Titan Tool Supply Company, Inc., 68 Comet Avenue, Buffalo, New York, 14216

Todd Research Industries,

Transidyne General Corp. 462 S. Wagner Road, Ann Arbor, Michigan 48106

Tremco Manufacturing Co., 10701 Shaker Blvd., Cleveland, Ohio 44104

UDT, (see United Detector Technology.)

UHU-Werk, H.u.M. Fischer G.m.b.H. 7580 Bühl, Baden

United Detector Technology, Inc., 1732-21st Street, Santa Monica, California 90404

V.A. & Co., Ltd., 88 Peterborough Road, London SW6, England

Van Waters and Rogers Scientific Inc. (VWR) (foreign and domestic), P.O. Box 3200, San Francisco, California 94119

Watkins and Doncaster, Four Throws, Hawkhurst, Kent, England

Wavetek, 9045 Balboa Avenue, San Diego, California 92123

Weiss & Son, Ltd., 17 Wigmore Street, London W1H ODN, England

W. Greenwood Electronic Ltd., 21 Germain Street, Chesham, Bucks, England

Wild Heerbrugg Ltd., CH-9435 Heerbrugg, Switzerland

Wilk Instruments, 789 S. Kellogg Avenue, Goleta, California 93017

William F. Nye, Inc., Box G 927, New Bedford, Massachusetts 02942

Winston, Electronics Co., P.O. Box 16156, San Francisco, California 94116 (dist. by David Kopf Instr.)

Wolff Industries, 2485 Huntington Drive, San Mareno, California 91108

W.-P. Instruments, Inc., P.O. Box 3110, 60 Fitch Street, New Haven, Connecticut 06515

Index